Design of Testable Logic Circuits

MICROELECTRONICS SYSTEMS DESIGN SERIES

Consulting editor   **G. Musgrave**
Brunel University

DESIGN OF
TESTABLE LOGIC CIRCUITS

R. G. BENNETTS

Cirrus Computers Limited

ADDISON-WESLEY PUBLISHING COMPANY
London · Reading, Massachusetts · Menlo Park, California · Amsterdam
Don Mills, Ontario · Manila · Singapore · Sydney · Tokyo

This book is dedicated to my colleagues at Cirrus Computers. If our earlier customers had read the book first, our jobs would have been considerably easier (but not as stimulating).

Prepared from camera-ready copy supplied by the author.

Printed in Finland by OTAVA. Member of Finnprint.

Cover design by Design Expo Limited.

Library of Congress Cataloguing in Publication Data

Bennetts, R. G.
    Design of testable logic circuits.

    (Microelectronics systems design series)
    Includes bibliographies and index.
    1. Logic circuits. 2. Logic circuits — Testing.
I. Title. II. Series.
TK7686. L6B45 1984   621.3819'535   84-11768

ISBN 0-201-14403-4

British Library Cataloguing in Publication Data

Bennetts, R. G.
    Design of testable logic circuits. — (Microelectronics systems design series)
    1. Logic circuits
I. Title. II.Series.
621.3815'37   TK7868.L6

ISBN 0-201-14403-4

BCDEF 898765

Foreword

For the past quinquennium the testing of electronic circuits has been an activity which happens at the end of a sequence of events listed as R and D, design, prototype, production. Certainly the emphasis has always been that testing is a post-design activity. In the past, this has been acceptable because the complexity of electronic circuits has been manageable, particularly from the point of view of 'observability' of component behaviour. Integrated circuit technology has changed that perspective. It is now imperative for the designer to consider testability at his early conceptual design stages, because the high costs of the inability to adequately test the complex components can cause total project collapse. Manufacturing industries at both the LSI/VLSI and systems levels are finding the high costs of testing (in some cases 60% of total costs) can only be reduced by bringing together design and test activities, leading to the concept of 'testable design'.

Dr. R.G. (Ben) Bennetts has researched in the area of test for digital circuits for over ten years and published many learned society papers. His earlier work was done as a faculty member of Southampton University where he developed his subject both in terms of research and teaching. He still teaches on the Brunel/ Southampton Universities' M.Sc. course in Microelectronics Design, although now he is a Director of Cirrus Computers Ltd. This company was initially involved in providing a test program service to the electronics industry. This experience was invaluable in establishing design criteria for testability and has led to Cirrus Computers developing computer aids to the electronics industry.

This experience of the author, both in the technical depth of the subject matter and as a lecturer, brings clarity to a text which helps bridge the gap between design and test both in practical and theoretical aspects. In the first chapter the overall scene is set in terms of the costs of testing and general strategies. The subsequent chapters deal with testability measures, with detailed examples of one particular schema. This provides the monitoring point for the other techniques such as scan design which are developed in detail, with worked examples throughout the text. This technique of worked examples is particularly valuable when applied to test generation algorithms such as PODEM (Path-Orientated DEcision Making) which hitherto have been the subject of institution papers and not tutorial texts for engineers and students. The fifth chapter brings the author's authority on the subject in complete focus with a series of guidelines for designing testable logic circuits. This is a text which will prove invaluable to all those who implement, manage or teach either design or test, but hopefully design *and* test. It is a pleasure to have this book in the Microelectronics Systems Design Series.

Brunel University,
Uxbridge, United Kingdom

G. Musgrave
June, 1983

Contents

Preface

Introduction

The purpose of this book is to present the tools and techniques of testable design of digital logic circuits. It is now acknowledged by the electronics manufacturing industry that testing is expensive and that 'testability' should form one of the criteria against which a design is judged. The question is – what does testability mean and how can the designer accommodate its requirements? Chapter 1 defines testability as follows.

> 'A circuit is testable if a set of test patterns can be generated, evaluated and applied in such a way as to satisfy pre-defined levels of performance, defined in terms of fault-detection, fault-location and test-application criteria, within a pre-defined cost budget and timescale.'

A less formal definition, attributed to Gordon Robinson of Cirrus Computers, is as follows.

> 'Testability is knowing how to do the things you need to do!'

Either way, testability is about influencing the design of a circuit so that the subsequent processes of testing become both manageable and soluble. This is the subject of the book.

About the book

The first chapter introduces the topic of digital testing and summarises the objectives, terminology, major test-programming activities and problems, including cost areas. The overall conclusion is that there is a need for testable design.

Chapter 2 presents the theory and practical application of one particular testability measure called CAMELOT (Computer Aided MEasure for LOgic Testability). The author has been involved in the development of this measure and the chapter discusses its application both in design (quantifying design changes, placing test points) and in developing test-generation strategies. The chapter concludes with a brief survey of other testability measures.

The important subject of scan design is covered in Chapter 3. This chapter presents the general philosophy of scan-path design and describes the IBM implementation known as LSSD (Level-Sensitive Scan Design). Also included in the chapter is a discussion on the generation and use of CRC (Cyclic-Redundancy Check) signatures leading to the concept of self test and built-in test and to the recent BILBO (Built-In Logic Block Observer) proposal.

Even if a circuit design is constrained to follow scan path, there is still the need to generate tests for the combinational section of the circuit. This problem has been well researched and many solutions proposed and implemented. Chapter 4 presents one particular approach based on a combination of deterministic and random path sensitisation. The method, known as PODEM/RAPS (Path Oriented DEcision Making/RAndom Path Sensitising), was developed originally within IBM to support the test-generation requirements of LSSD circuits. PODEM/RAPS is epitomized by simplicity and efficiency.

Practical guidelines have always had an important role to play in designing testable logic circuits and Chapter 5 presents a comprehensive checklist of 'do's and don'ts'. The chapter is an extended and revised version of part of Chapter 9 in the author's earlier book on digital testing:

R.G.Bennetts, 1982, 'Introduction to digital board testing' Crane-Russak (New York), Edward Arnold (London).

The Appendix contains a reprint of the data sheet for a recent (1983) integrated circuit device from Advanced Micro Devices, Inc. This device, the Am29818 Serial Shadow Register, contains both an operational register and a shadow scan-path register. Together with its Monolithic Memories counterpart, the SN54/74S818, the devices are the first devices available with scan-path facilities for use by any board or system logic designer. As such, they represent a significant breakthrough in terms of making testable design techniques widely available. Undoubtedly other devices such as these will follow.

Finally, it should be stated that the material in Chapters 2, 3 and 4 have not appeared in book form before. Also, each chapter is designed to be relatively free standing and accompanied by a comprehensive list of references.

Who should read the book?

The book will have value to the following sections of the electronics engineering community.

(a) Students of digital electronics engineering. These people will shortly become designers or test engineers and an early appreciation of design for testability will help to reduce the overall costs of testing future designs.

(b) Practising digital design engineers. Designers are continually learning and applying new skills. Design for testability is a skill that should be a standard part of any logic designer's repertoire. If the skill is missing, which is often the case, the book will serve as a useful place to start the educational process.

(c) Practising test engineers. Design for testability is central to the digital test-engineering community, i.e. those who design, manufacture and make use of digital test systems. An awareness of the disciplines of design for testability is vital if communication paths between designers and test engineers are to be kept open.

(d) Managers of designers and test engineers. Design for testability will only work in practice if there is an understanding of a commitment to the need by those who manage design and test activities. Such people should also read the book but may chose to omit detailed descriptions. As a guide, the author would recommend Chapter 1, Chapter 2 (Sections 2.1, 2.5), Chapter 3 (Sections 3.1, 3.4) and Chapter 5 (top-level guidelines only).

(e) Teachers of both digital design and test topics. The material in this book should form part of any higher-educational digital electronics course that goes beyond the basic principles of logical design.

Acknowledgements

The following acknowledgements to organisations and to people are due.

CAMELOT was developed originally by Cirrus Computers under a development contract placed by British Telecom Research Laboratories (UK). The program is now owned by BTRL and CAMELOT is a registered trademark. Cirrus Computers has the right to market the program under the terms of a licence agreement with British Telecom.

Other acknowledgements to organisations include the Institute of Electrical Engineers (parts of Chapter 2 on CAMELOT are based on an earlier paper written by the author and published by the IEE); Crane-Russak, Inc., (for permissions to modify, extend and re-publish the list of guidelines in Chapter 5); Advanced Micro Devices, Inc., (for permission to reproduce parts of the Am29818 data sheet in the Appendix).

In terms of people, I would like to acknowledge the efforts of the following individuals: John Anderson, Michael Bending, John Ibbotson, David Potts and Gordon Robinson of Cirrus Computers, and Colin Maunder of British Telecom for critically and constructively contributing to a review of an earlier version of the manuscript;

Gerry Musgrave (Series Editor) for inviting me to contribute to the Microelectronics Systems Design Series; Jean Staley for her patient work in creating and continuously editing the manuscript, this time right through to the camera-copy; and lastly, my wife Carol and children Mark, Kevin and Helen who must have wondered whether I would ever finish writing the book. To all these people I offer my thanks.

R.G.Bennetts
Cirrus Computers
June, 1983

1

Digital testing:
the need for testable designs

1.1   INTRODUCTION

Testing digital electronic systems at chip, board or system level
is expensive.   As a percentage of the total life-cycle cost of a
product, estimates of the costs of testing vary from very low
(less than 10%) to very high (greater than 60%) according to the
position on the life-cycle and who is having to find the money.
Traditionally, the two disciplines of logic design and test
programming have been considered to be separate; design preceded
test and designers were not involved in the test process.
    Recently however, the increasing complexity of the design
primitives has created a situation where it is now extremely easy
for a digital circuit designer to produce a circuit design which
is virtually untestable in real terms. (The term 'testable' will
be defined later in Section 4.1 of this chapter.) This situation
applies to all levels of design, ranging from single chip through
to a full digital system based on a collection of printed-circuit
boards   (PCBs).   It   is   now acknowledged by designers and
test-programmers alike that test requirements must influence the
design activity if final designs are to be testable. This
situation realisation has given rise to a number of ways for
improving the testability of a design - some formal, some less
formal - and this is the subject of this book.
    The purpose of this chapter is to summarise the activities, to
review the status of digital testing, and to establish the need
for testable designs.   The remaining chapters will   focus on
specific ways for satisfying this need. The reader who wishes to
pursue further the summarised material of Chapter 1 is referred to
the author's earlier book, reference 1.1 in the bibliography at
the end of the chapter.

1.2 DIGITAL TESTING: OBJECTIVES AND TERMINOLOGY

The primary objective of testing digital circuits at chip, board
or system level is to detect the presence of hardware failures
induced by faults in the manufacturing processes or by operating
stress or wearout mechanisms.   Such testing is referred to as

'go/no go' testing.

If the circuit or system is designed to be repairable, the secondary objective of testing is to locate the cause of a fault with enough precision and correctness to enable an effective repair to be carried out. This form of testing is called 'diagnostic' and implies both detection and location. A 'fault' relates to the physical failure mechanism, e.g. an open-circuit track, whereas 'fault-effect' relates to the logical effect of the fault on a signal-carrying node e.g. the node is stuck at logic 1. The behaviour of most faults is modelled adequately by either the stuck-at-1 (s-a-1) or the stuck-at-0 (s-a-0) fault-effect models or by the wired-AND, wired-OR bridging fault-effect models.

In general, a digital circuit is an assembly of basic logic gates, flip-flops, and more complex digital devices such as shift registers, counters, ROMs, RAMs, microprocessors, and microprocessor-support circuits. The circuit may also contain other electronic devices such as transistors, operational amplifiers, pull-up or pull-down resistors and decoupling capacitors. Normally, however, a circuit can only be driven (stimulated) by a tester through certain access points (chip pins or board edge-connector fingers). Similarly, the circuit can usually only be sensed (monitored) at other defined access points. Inputs that can be driven and outputs that can be sensed are called 'primary inputs' (PI) and 'primary outputs' (PO) respectively. The word 'test' or 'test pattern' means a specified PI stimulus plus the expected fault-free PO response. The term 'fault-cover' refers to the set of faults detected either by an individual test or by a set of tests, in which case the fault-cover is said to be cumulative.

Tests are applied to digital circuits through a variety of programmable Automatic Test Equipments (ATE). Chief amongst these are device, bare-board, in-circuit, functional and field-service testers. Each tester is designed to carry out a specific testing task associated with the manufacturing cycle or field-use of the product. Modern ATEs are based on computer systems programmed to interface with the device-under-test (DUT) or board-under-test (BUT) through a custom-designed driver/sensor system. For PCB functional testers, this system may contain a roving sensor, more-commonly called a guided probe, for use as an additional aid to fault location.

1.3 TEST-PROGRAMMING ACTIVITIES AND PROBLEMS

It is convenient to partition test programming into three major activities: test generation, test evaluation and test application. Test generation is the process of providing a set of test input stimuli plus expected fault-free responses to meet the requirements of a target fault list. The generation of tests is subject to constraints imposed both by the tester and by the circuit itself in terms of access. Test evaluation is the process of quantifying just how good the set of tests is against the target fault list, usually specified in terms of single s-a-1, s-a-0 faults on all signal-carrying nodes in the circuit. Test application is the physical process of applying the tests to a real circuit. A summary of the problems currently encountered in each of these areas follows.

1.3.1 Test generation

Tests for logic circuits are generated to satisfy one or more of a variety of specific objectives: change the value of a circuit node; set up a sensitive path for a specific fault-effect (e.g. node s-a-1, s-a-0); exercise a recognisable sub-circuit through all of its states; and so on. Procedures for achieving these objectives are largely based on programmed techniques to generate sensitive paths, coupled with user interaction to establish the test strategy and to solve the more complex problems.

There are two factors that affect the ease, or otherwise, of generating test patterns for a digital circuit. They are both related to the tester's visibility of the circuit elements and resolve into controllability (CY) and observability (OY). The more controllability and observability features are built into a circuit, the easier it becomes to create the right circumstances to excite and propagate fault conditions. For PCBs based on the use of SSI and MSI devices the ability of logic designers to produce difficult-to-test circuits is well known. More recently, designers of circuits based on gate-array devices have also demonstrated this ability! The problem arises from the use of stored-state devices within complex feedback structures. The more complex the structure, the more difficult it becomes both to control and to observe the behaviour of the circuit. For bus-structured designs, the test-generation problems are more concerned with the complexity of the actual LSI and VLSI devices than with controllability and observability access. Specifically, problems arise from:

(a) ambiguous, incomplete or incorrect data sheets;
(b) differences between prime-source and second-source devices;
(c) the variety, complexity and rate of introduction of new devices;
(d) the volume of test data required to test complex devices, especially RAM and microprocessor devices;
(e) lack of knowledge about the precise way in which the device can fail.

1.3.2 Test evaluation

The evaluation of test patterns is carried out against the target fault list. It requires either a known-good-device (KGD) or known-good-board (KGB) on which faults can be physically inserted to measure the performance of the test program. Alternatively, a logic fault simulator can be used to assess fault-cover performance through insertion of faults into a software model of the circuit. Physical fault insertion is limited in its capability either because access is limited (as with an integrated-circuit device) or because the inserted faults have to be restricted to those that do not cause permanent damage to the circuit. For example, the output of a TTL gate can be grounded to

4

emulate the effect of a s-a-0 fault, but it cannot be connected to the +5v rail, either directly or through a pull-up resistor, to emulate the s-a-1 fault. To do so for any significant length of time would cause irreversible damage to the device.

The only practical solution to the test-evaluation requirement is to make use of a logic fault simulator. Here the basic strategy is to exercise a model of the circuit and assess both its fault-free and faulty behaviour under application of the test patterns. Faults are taken from the target fault list and inserted into the model either one at a time (serial fault simulator) or in predefined groups (parallel fault simulator). Alternatively, faults that are ultimately detected at the POs are deduced as the fault-free values are determined (deductive or concurrent fault simulator).

In all cases, the fault simulator requires some form of circuit model to allow assessment both of the fault-free behaviour of the circuit and of the behaviour in the presence of a fault. ror LSI/VLSI devices and PCBs using these devices, the difficulty is to produce a model to represent this behaviour accurately – largely for the same reasons as outlined in the previous section (inadequate data sheets, device differences, device complexity, and unknown failure mechanisms).

1.3.3 Test application

Test application brings the circuit into contact with the ATE. Problems here are mostly concerned with the limitations of the ATE in terms of hardware limits (maximum test-application rates, limited driver/sensor change sequences, etc.) and restrictions on facilities (fault-dictionary with or without guided probe, pulse-catching capability, etc.). The fault-location problem can be particularly difficult to solve for circuits with global feedback structures if the cause-effect relationship around a closed loop cannot be resolved (the so-called 'loop-breaking' problem). Physical access and interface requirements can also cause problems if devices on boards are physically close to each other or if the test program requires additional access through non-standard flying leads.

1.3.4 Major cost areas

To summarise, the major costs associated with the various testing activities are as follows:

For test generation:

(a) the actual process of generating new test programs for new devices or PCBs, carried out manually or with some computer assistance;
(b) development or procurement cost of general-purpose test-generation programs, based largely on classical sensitive-path algorithms;
(c) procurement of computer hardware on which to run the test-generation software;

(d) actual cost of generating test programs for new device or new board designs against pre-specified acceptance standards;
(e) support and maintenance of items (a) to (d) above;
(f) availability and continual training of skilled test-programming engineers.

For test evaluation:

(a) development or procurement cost of general-purpose fault-simulator software plus, as above, associated computer hardware;
(b) costs associated with the continual development of suitable device models for use within the fault simulator;
(c) cost of fault-simulation run times – usually considered a major contributor to overall costs;
(d) support and maintenance of items (a) to (c) above;
(e) training.

For test application:

(a) ATE procurement, support, maintenance and repair costs;
(b) costs associated with the interface requirements for each circuit type to be tested;
(c) costs associated with testing throughput: time to set up the circuit on the ATE, average time to detect a fault, average time to repair and retest;
(d) effect of incorrect diagnosis in terms of the cost of replacing non-faulty components and then retesting;
(e) effect of inadequate fault cover in terms of the cost of shipping a faulty product to the customer;
(f) training.

In general, the total life-cycle cost of testing will be some combination of all these costs and will vary according to the following factors:

(a) expected yield and quality level requirement;
(b) life-cycle test strategy in terms of ATE mix and number of testers per type;
(c) volume and variety of different designs;
(d) centralised or de-centralised field servicing and repair strategy.

The general subject of ATE economics and life-cycle costs is not pursued in this book. The reader interested in pursuing this topic further is referred to B. Davis' book, reference 1.2 in the bibliography at the end of this chapter.

1.4   NEED FOR TESTABLE DESIGN

It is not possible to quantify or apportion any of the factors listed in the previous section without a detailed statement of the test requirements and environment. What is clear however is that the total costs associated with the life-cycle testing of a product are now generally considered to be excessive, and likely to continue to be so, by many sections of the electronics manufacturing industry, ranging from chip manufacturers right through to full system manufacturers. As a result, there has been considerable activity aimed at reducing specific areas of these costs. This has given rise to the requirement for 'Design For Test' and certain methods have emerged as both viable and durable. Before commenting on these methods, we should define what we mean by 'testable'.

1.4.1 Definition of testability

Testability relates to cost, as discussed in Section 1.3.4. An informal definition of testability is as follows:

> A circuit is 'testable' if a set of test patterns can be generated, evaluated and applied in such a way as to satisfy pre-defined levels of performance, defined in terms of fault-detection, fault-location, and test application criteria, within a pre-defined cost budget and timescale.

The key factor in this definition is the need to satisfy certain technical requirements within a budgeted allowance. If the actual cost of testing is greater than the pre-defined cost, we can conclude that the design was originally untestable by the definition above. The remedy to make the design testable is either to increase the budget allowance or to reduce the cost of one or more of the major contributing cost factors. If we adopt the second course of action, any procedure for lowering testing costs can be considered to be a Design-For-Test procedure.

1.4.2 Design-For-Test techniques

The remaining chapters of this book describe three major approaches for reducing testing costs by designing testable logic circuits. The approaches can be categorised as follows:

(a) numerical assessment of the controllability and observability features of a circuit design, leading to a measurement of the circuit's testability which can be used during the design stage (Chapter 2);

(b) techniques for designing highly-structured and, in some cases, self-testing circuits containing scan-path facilities aimed at improving access to and from the circuit (Chapter 3), plus effective techniques for generating tests for scan-designed circuits (Chapter 4);

(c) checklist of practical guidelines aimed at reducing both test-generation and test-application costs (Chapter 5).

1.4.3 Benefits and commitments

The benefits of Design–For–Test can be summed up in a single phrase – reduction in overall design cycle times and test costs without sacrifice to the quality of the product. Additionally, individual testing activities become manageable and designers become educated into the ways of testing. Only in this way will the activities of design and test become truly integrated. But, a Design–For–Test policy will only work if there is a firm commitment by the designers themselves and also by the managers of those who design, those who manufacture, and those who test. Failure to recognise the need for this commitment will inevitably lead to failure of the policy thereby leading to higher manufacturing and field servicing costs with, at the very least, a corresponding reduction in profit margins and product reliability and credibility.

BIBLIOGRAPHY

1.1 Bennetts R.G., 1982, "Introduction to digital board testing", Crane–Russak (New York), Edward Arnold (London).

1.2 Davis B., 1982, "The economics of automatic test", McGraw–Hill (London).

2

Testability measurement: the CAMELOT approach

2.1   QUANTIFYING TESTABILITY

There are two ways of quantifying the testability of a circuit design without actually completing the tasks of test generation and evaluation. These are referred to as the 'scoring' and 'algorithmic' methods.   Scoring methods operate by identifying circuit features which either contribute to or detract from testability.   Each feature is given a number of points representing the magnitude of the effect its presence will have on testability.   For example, the absence of a simple means of initialising a circuit may severely degrade its testability, but the effect of a wired–AND structure may not be so critical. Once the number of occurrences of each feature has been established, the total number of points for and against a design can be calculated.   These two scores can then be combined to produce a single testability rating.

Scoring systems have two principal advantages.   First, they are easy to implement, often not requiring the use of a computer. Second, they allow assessment of the effect on testability of the various mechanical aspects of a finished design, e.g. component separation, connector location, etc.   However, scoring is essentially a coarse measure of testability. The main value of scoring is that it allows a testable design to be distinguished from one that, on average, is less testable. It does this by providing the designer with a comprehensive checklist of design features which the designer should either avoid because they create test problems, or include because they simplify test solutions.   Chapter 5 presents and comments on such a checklist, but the items are not rated in terms of their relative value.

The algorithmic measures are computer based and produce testability estimates by analysing a topological description of the circuit. The measures have the advantage of quantifying testability for each circuit node, thus allowing the construction of a testability profile of the circuit, usually in the form of a histogram distribution. Comparison of the various nodal testability values allows the areas of poor testability to be readily identified and the relative merits of various improvement techniques to be assessed.

Measures that fall into this category are largely based on mathematical models of the two tasks inherent in the test generation process. The tasks are described with the aid of Fig. 2.1.

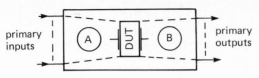

Fig. 2.1 Testing process.

A general purpose strategy for producing test patterns for a complex digital circuit is to attempt to apply a functional test to each recognisable subfunction or device within the circuit. For example, a counter should be made to count, a shift register to shift, etc. In order to apply such a test to a particular subfunction, it is first necessary to establish a sequence of logic patterns at its input. To produce each pattern, it is necessary to establish fixed values on the nodes in region A (Fig .2.1) of the circuit by setting values on the circuit's primary inputs. The ease of completing this task at each node is defined to be its 'controllability'.

Having completed the first task, the response of the subfunction must be made observable at the circuit's primary outputs. This will require sensitive paths to be set up through region B and will involve establishing fixed values on other nodes throughout the circuit using the assigned and unassigned primary inputs. The ease of completing this process for each node is defined to be its 'observability'.

The testability (TY) of each node is, then, a function of its controllability and observability values.

The remainder of this chapter describes in some detail the theory and use of one particular testability analysis system, known as CAMELOT (Computer-Aided MEasure for LOgic Testability), developed originally by the author and his colleagues. Other measures have also been developed however, and these are mentioned briefly at the end of the chapter.

2.2 CONCEPTS OF CONTROLLABILITY

2.2.1 Definition and transfer of controllability values

In CAMELOT, controllability values (henceforth denoted CY values) are constrained to be in the range 0 to 1. The maximum value of 1 represents a node, such as a primary input, where it is as easy to establish a logic 1 as it is a logic 0. At the opposite extreme, a CY of 0 represents a node which cannot be set to one or other of the two logic states. In practice, the majority of the nodes in a circuit will have CYs between these two limits.

Having defined the CY values of primary inputs, we must now develop a theory for the transfer of CYs across devices in order to calculate values at other nodes.

Consider a device such as is shown in Fig. 2.2. If the inputs to the device were directly controllable, then the CYs of its outputs would simply reflect the relative ease with which each output could be set to 0 and to 1, as determined by the logical transfer function of the device. In the general case, however, the inputs to the device will not be 100% controllable. The output CYs of the device must therefore take into account both the ease of transfer across the device and the CY values on the inputs. The expression used to calculate each output CY is thus:

$$CY(\text{output node}) = CTF \times f\{CYs(\text{input nodes})\} \qquad (2.1)$$

where the CTF is the Controllability Transfer Factor of the device for the output concerned and the function f combines the CYs on all the inputs to the device that are able to control the particular output. The CTF and function f are defined in the following sections.

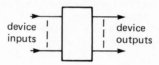

Fig. 2.2 Typical device.

2.2.2 Controllability Transfer Factor (CTF)

The CTF of an output is a measure of the degree of departure from the situation in which it is as easy to generate an output value of 1 as it is a value of 0. The CTF is dependent only on the logic function of the device and not on the position of the device in the circuit.

A simple, but effective, way of quantifying the CTF is given by:

$$CTF = 1 - \left| \frac{N(0) - N(1)}{N(0) + N(1)} \right| \qquad (2.2)$$

where $N(0)$ is the total number of ways that a 0 can be produced on the device output and $N(1)$ is the total number of ways of producing a 1. Thus the CTF is 1 in the case where $N(0)$ and $N(1)$ are equal (as, for example, for an exclusive-OR gate). In the unlikely event of $N(0)$ or $N(1)$ being 0 then the CTF would be 0, indicating that there is no control of the output state. Generally, $0 < CTF < 1.0$.

Figure 2.3 contains examples of CTF calculations for simple devices. For these devices, as for all combinational devices, the values of $N(0)$ and $N(1)$ can be obtained from the device's truth table. Note that, for devices with several outputs, each output will have its own CTF value and, in the general case, these values will not be the same.

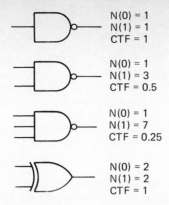

$N(0) = 1$
$N(1) = 1$
$CTF = 1$

$N(0) = 1$
$N(1) = 3$
$CTF = 0.5$

$N(0) = 1$
$N(1) = 7$
$CTF = 0.25$

$N(0) = 2$
$N(1) = 2$
$CTF = 1$

Fig. 2.3 Some CTF calculations for simple functions.

The procedure for calculating the CTFs for stored-state devices is slightly more complex. Consider, for example, a J-K flip-flop, such as shown in Fig. 2.4.

$$Q^+ = [(J\overline{Q} + \overline{K}Q) T + Q\overline{T}] PC + \overline{P}$$

$$\overline{Q}^+ = [(\overline{J}\overline{Q} + KQ) T + \overline{Q}\overline{T}] PC + \overline{C}$$

(T = 1 signifies the presence of a negative clock edge)

Fig. 2.4 SN7476 flip-flop.

The Boolean expressions given for the Q and $\overline{Q}$ outputs have been derived from the truth table for the device, and account both for clocked and unclocked behaviour. Analysis of these expressions will reveal the number of ways of setting Q ($\overline{Q}$) to 1 and, hence, the number of ways of setting it to 0. This is done in the following way:

Consider the expression for Q+:

$$\begin{aligned} Q+ &= [(J\overline{Q} + \overline{K}Q)T + Q\overline{T}]PC + \overline{P} \\ &= J\overline{Q}TPC + \overline{K}QTPC + Q\overline{T}PC + \overline{P} \quad \text{(by expansion)} \quad (2.3) \end{aligned}$$

Each conjunctive (logical product) term in the expanded form of this expression represents a collection of 1-entries on a Karnaugh map for the output Q+. If it is known how many 1-entries each conjunctive term represents and it can be guaranteed that no conjunctive term overlaps another (i.e. the set of conjunctive terms is disjoint) then N(1) can be calculated.

Satisfying the first condition is easy. If Q+ is a function of n variables and the ith conjunctive term contains j variables (j ⩽ n), then the number of 1-entries represented by the ith term is given by $2^{(n-j)}$. The second condition is not quite so easy to satisfy, but a simple procedure does exist which can be used in this application. The procedure is based on an exhaustive application of two basic processes:

(a) a test for disjointness between pairs of conjunctive terms;
(b) if the test fails, the modification of one term relative to the other.

The test for disjointness is performed by checking each pair of terms in a systematic but exhaustive fashion to see if they contain at least one common variable which is present in its complemented form in one term and its uncomplemented form in the other. If this condition is true, then the two terms map into different domains of the Karnaugh map and cannot overlap, i.e. they are disjoint. If the test fails, then the variables present in one term but not in the other are reintroduced into one of the terms in such a way as not to modify the function (i.e. the total number of 1-entries produced by the two terms) but to render them disjoint. A suitable procedure is as follows.

PROCEDURE: Let T1 and T2 be two non-disjoint product terms in a Boolean sum-of-products expression, and let RC = T1\T2 be the relative complement* of T1 and T2. Define RC by the non-empty set {X1,X2,...,Xn}. (All Xs are written in uncomplemented form but this does not affect the generality of the result.) The procedure by which T1 and T2 are converted into a disjoint collection of terms is described by the following expansion:

$$T1 + T2 = T1 + \{X1 + X1.\overline{X2} + X1.X2.\overline{X3}.+....+(X1.X2.X3....\overline{Xn})\}T2$$

$$(2.4)$$

Effectively, the expansion is based on a controlled reintroduction of missing variables to refine the collection of 1-entries represented by a particular term. By inspection, each pair of terms in the right-hand-side of expression 2.4 is disjoint.

Returning to expression 2.3 we see that each term is already disjoint from all the others. The total number of 1-entries is therefore:

$\overline{J}QTPC$	representing	2	1-entries (variable K missing)
$\overline{K}QTPC$	representing	2	1-entries (variable J missing)
$QTPC$	representing	4	1-entries (J and K both missing)
P	representing	32	1-entries (J, K, T, Q, C all missing)

* The relative complement, T1\T2, is the set of items present in T1 but not present in T2. Hence if T1 = {a,b,c,d} and T2 = {b,c,e}, then T1\T2 = {a,d}. Note that the relative complement is non-commutative i.e. T1\T2 is not equal to T2\T1.

giving a total of 40 1-entries.

Before we conclude that N(1) is 40, however, we should take note of another factor, namely invalid or transient states. For the device of Fig. 2.4, the following input combinations do not produce a stable output:

P = 0, C = 1, Q = 0   representing 8 invalid states
P = 1, C = 0, Q = 1   representing 8 invalid states
P = 0, C = 0, Q = 0   representing 8 invalid states

If, for any reason, any of these 24 sets of conditions should occur, then the initial value of Q as defined in each set is unstable and will eventually reach a stable state opposite to the one specified. We should take care to remove any such invalid states from the 1-entries derived above. In this example, the first three 1-entry terms all specify valid conditions. However, some 24 of the Karnaugh map entries are invalid and the number of 1-entries must be reduced accordingly. In the case of the fourth term, $\overline{P}$, the reduction is by 16 corresponding to the 16 P=0 invalid states, leaving 16 valid 1-entries. The total value for N(1) is therefore 24. With a knowledge of N(1), a value is obtained for N(0) by computing the remaining number of valid states for the device. In this case, there are 2^6 possible states of the device, some of which are invalid, so N(0) is given by:

$$N(0) = [2^6 - \text{total invalid states}] - N(1) = 16$$

This produces a CTF value of 0.8 for the Q output, which, by symmetry is also the value obtained for the $\overline{Q}$ output.

One final comment should be made on this calculation. The CTF has been calculated on the basis of Q+ being a function of five inputs to the device (J,K,T,P,C) plus the present value of the stored state Q. Strictly speaking, Q is not an input to the device but the inclusion of it as an input variable accounts for the fact that the controllability of the future state of the output Q+ is dependent on its present state.

2.2.3 Computing output CY values

Consider a device with n inputs and m outputs. In general, each output of the device is a function of some, or all, of the device inputs and its CY can be quite different from that of neighbour outputs. This suggests that each output has its own CTF (as discussed previously) together with a list of inputs that control its value. Information on these two properties must be held in a device library.

The CY of a particular output Z1, for example, is given (equation 2.1) by:

$$CY(Z1) = CTF(Z1) \times f[CY(\text{input nodes controlling } Z1)] \qquad (2.5)$$

What is the function f?

Initially, the choice would seem to be between the arithmetic and geometric means of the CYs of the inputs that can affect the output, Z1. However, the geometric mean can be ruled out because

$$\{CY(X1) \times CY(X2) \times ... \times CY(Xn)\}^{1/n}$$

equals 0 if any of the input CYs are 0. Figure 2.5 illustrates a situation in which one of the device inputs has zero controllability and yet, clearly, the output is controllable.

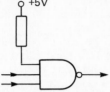

Fig. 2.5 Zero input controllability.

The arithmetic mean, however, does cater for the possibility of zero CYs, but it also has limitations. To understand this, consider the D-type flip-flop shown in Fig. 2.6.

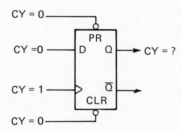

Fig. 2.6 SN7474 flip-flop.

Using a simple arithmetic mean for the function f would produce the following equation for the CY of output Q of the flip-flop:

$$CY(Q) = CTF(Q) \times \{CY(PR) + CY(D) + CY(CLK) + CY(CLR)\}/4 \quad (2.6)$$

If $CY(CLK) = 1$ but $CY(PR) = CY(CLR) = 0$ because they are held at their enable level and $CY(D) = 0$ because the state of D is fixed, then equation 2.6 produces:

$$CY(Q) = CTF(Q) \times 0.25$$

This result is obviously incorrect, because the output value on Q cannot be altered. Its CY should therefore be zero.

This suggests a modification of the arithmetic mean function to take account of the two separate modes of behaviour of most stored-state devices – the clocked and unclocked modes. For the unclocked mode, a simple arithmetic mean of the CYs of the asynchronous inputs is satisfactory. In the clocked mode, the CY of each of the clock-controlled inputs must be multiplied by the CY of the clock signal before forming the mean. In the case of multiple clock signals, the CY of each clock-controlled input must be multiplied by the product of the multiple clock CY values. The general expression accounting for both modes of operation (asynchronous and synchronous) and for multiple clocks is therefore:

$$CY(output) = CTF(output) \times f(input\ CYs),$$

where

$$f(input\ CYs) = \frac{1}{(i+j)}\ [\ \Sigma\ (CY: i\ async\ inputs)$$
$$+ \left\{\ \Pi(CY: clock) \times \Sigma\ (CY: j\ sync\ inputs\ )\right\}\ ] \qquad (2.7)$$

Thus, for the D-type flip-flop considered earlier, equation 2.7 becomes:

$$CY(Q) = CTF(Q) \times \{CY(CLR) + CY(PR) + CY(CLK).CY(D)\}/(3)$$

$$= 0\ if\ CY(PR) = CY(CLR) = CY(D) = 0,\ CY(CLK) = 1$$

Notice that if the PR and CLR inputs were tied off (held at logic 1), then $CY(PR) = CY(CLR) = 0$ and the expression for $CY(Q)$ becomes:

$$CY(Q) = CTF(Q) \times \{CY(CLK).CY(D)\}/(3)$$

If $CY(CLK)$ and $CY(D)$ were both 1, this revised expression would produce:

$$CY(Q) = CTF(Q)/(3)$$

The division by 3 is unnecessary in this case and makes the result incorrect. This type of problem points to the fact that, in practice, CTF values and CY transfer equations must be based on the implemented function. In the case of the flip-flop with PR and CLR tied off, the correct approach is to ignore the PR and CLR altogether, i.e. to interpret equation 2.7 as:

$$CY(Q) = CTF(Q) \times \{CY(CLK).CY(D)\}$$

2.2.4 Controllability transfer around feedback paths

The presence of feedback external to a device (global feedback) creates a problem when calculating nodal CY values throughout a circuit. A set of simultaneous equations is created and these must be solved. Consider the circuit shown in Fig. 2.7. For this circuit:

CY(U1.1) = 1.0
CY(U1.2) = X   (initially unknown)
CY(U1.4) = Y   (initially unknown)
CY(U1.5) = 1.0

The transfer factor for both devices is 0.5, so:

CY(U1.3) = Y
$\quad\quad\quad$ = 0.5 x (1 + X)/2

CY(U1.6) = X
$\quad\quad\quad$ = 0.5 x (1 + Y)/2

Solving these equations produces the value 0.33 for both X and Y.

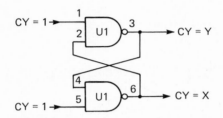

Fig. 2.7 Simple latch circuit.

2.3 CONCEPTS OF OBSERVABILITY

2.3.1 Definition and transfer of observability values

We will now consider the second factor in testability – observability (henceforth denoted OY). Returning to Fig. 2.1, we see that the major objective in region B is to propagate fault-effect information from the input side of a device to its output – generally, from a particular input to a particular output. Accordingly, the OY of a node is defined to be a measure of the ease with which the value of the node can be observed at one or more of the primary outputs of the circuit. This means that the OY of a primary output is 1. Also, the OY of a node at the node itself is 1; this value decreases as the point of observation passes along the sensitive path towards a primary output.

The amount by which the OY value is reduced from the input to the output of a device on the sensitive path is, in the case of a device whose other inputs are perfectly controllable, determined by its Observability Transfer Factor (OTF). This factor reflects the ease of propagating a change in value on one input of the device to a particular output.

In general, however, the propagation of fault-effect data through a device is a function both of establishing sensitivity on a particular input and of establishing fixed values on some, or all, of the other device inputs to allow transfer of sensitivity to the particular output. The ease of satisfying the second of these requirements is obviously a function of the CYs on those other inputs, this situation being shown in Fig. 2.8. Therefore:

OY(at output) = OTF x OY(at input) x g(CYs on supporting inputs)
(2.8)

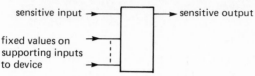

Fig. 2.8 Sensitivity transfer.

2.3.2 Observability Transfer Factor (OTF)

Intuitively, the OTF from input I of a device to its output O, denoted OTF(I–O), should be 0 if there is no way of propagating fault–effect data between the two points. Conversely, the OTF should be 1 if propagation always takes place, regardless of the values on the supporting inputs. Generally, however, the OTF will lie between these extremes.

A formal way of defining sensitivity transfer is contained in the concepts of the D–algorithm: D–algorithm in particular, in the concepts of propagating and non–propagating D–cubes (PDCs and NPDCs respectively). Each PDC identifies the sensitive path input, the fixed value combinations that support the path and the sensitive path output. Each NPDC, on the other hand, identifies the sensitive path input, the fixed value combinations that block the path and the insensitive output. The total number of distinct, but unpolarised* PDCs for an input–output pair (I–O) quantifies the number of possible ways of propagating fault–effect data. We will denote this number by N(PDC:I–O).

Similarly, the total number of distinct but unpolarised NPDCs from I to O, denoted N(NPDC:I–O), indicates the number of ways of blocking the transfer of sensitivity across the device. The ease, or otherwise, of being able to make the sensitivity transfer, the device's OTF, can therefore be measured by the ratio:

$$OTF(I-O) = \frac{N(PDC:\ I-O)}{N(PDC:\ I-O) + N(NPDC:\ I-O)} \qquad (2.9)$$

Note that each PDC usually represents two sensitive paths through the device because of its polarised expansions. Later on, though, it is shown that this is not always true and that care is needed

* In the case of a 2–input NAND gate, the D–cubes (D1/$\bar{D}$) and ($\bar{D}$1/D) are both valid PDCs. They are polarised in the sense that use is made of the invert function on the D variable. However, the two cubes represent the same path through the device and this single path option could be denoted simply (D1/D), i.e. the D is used only to indicate sensitivity status. This is the unpolarised form of the D–cube.

when determining the exact numbers of PDCs and NPDCs for a device.
An alternative, but equivalent, form of equation 2.9 is given by:

$$\text{OTF (I--O)} = \frac{N(\text{SP: I--O})}{N(\text{SP: I--O}) + N(\text{IP: I--O})} \qquad (2.10)$$

where $N(\text{SP:I--O})$ is the total number of different sensitive paths from I to O and $N(\text{IP:I--O})$ is the total number of insensitive paths. Where there is a possibility of confusion, it is better to use equation 2.10 as the more basic form of the definition. The reason for this is clarified later when we consider the calculation of OTF values for stored-state devices.

There is one further observation that should be made at this stage. Transfer of sensitivity through a device is assumed to occur from a single input to the relevant output. $N(\text{PDC:I--O})$ really corresponds to the number of single-D propagating D-cubes for the device. Multiple sensitive paths can also be created through devices, as defined by the multiple-D propagating D-cubes for the device (for example, $D\overline{D}/\overline{D}$ for the 2-input NAND gate). In general, however, the creation of multiple-D sensitive paths is avoided when generating tests, unless there is no alternative single-D option, and the extra work involved in defining and handling OTFs for multiple-input paths is not considered worthwhile.

Some OTF calculations for simple devices are shown in Fig. 2.9.

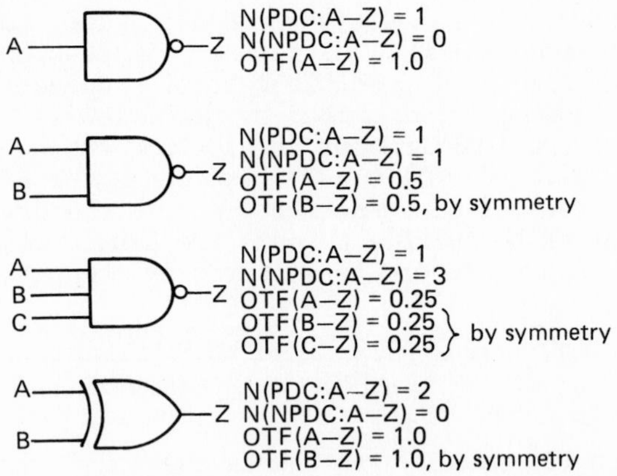

Fig. 2.9 Some OTF calculations for simple functions.

The method of calculating OTFs for more complex stored-state devices is described with the aid of Fig. 2.10. This figure illustrates a J-K flip-flop with Clear, but no Preset; for example the SN7473.

One method of determining the number of ways of sensitising an input-output path is to list all the possibilities and examine each one in turn. In effect, this is the fully exhaustive 'brute force' technique. The result of performing this process for the flip-flop is shown in Fig. 2.11.

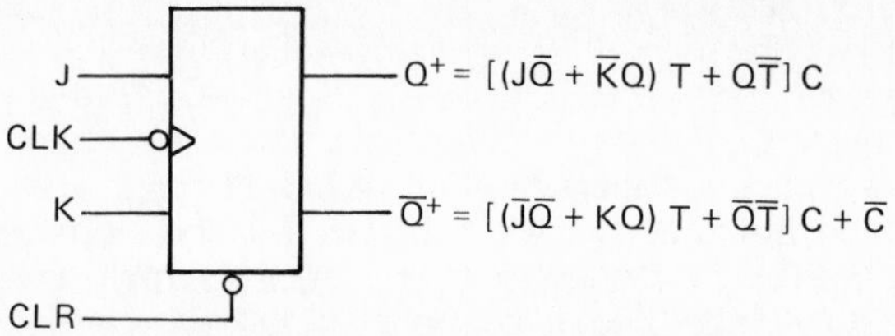

$$Q^+ = [(J\bar{Q} + \bar{K}Q)\, T + Q\bar{T}]\, C$$

$$\bar{Q}^+ = [(\bar{J}\bar{Q} + KQ)\, T + \bar{Q}\bar{T}]\, C + \bar{C}$$

Fig. 2.10 SN7473 flip-flop.

Initial state					Fault-free response	Fault on ? input detected at Q output			
J	K	T	C	Q		J input	K input	T input	C input
0	0	0	0	0	no change	N	N	N	N
0	0	0	0	1	*	*	*	*	*
0	0	0	1	0	no change	N	N	N	N
0	0	0	1	1	no change	N	N	N	Y
0	0	1	0	0	no change	N	N	N	N
0	0	1	0	1	*	*	*	*	*
0	0	1	1	0	no change	Y	N	N	N
0	0	1	1	1	no change	N	Y	N	Y
0	1	0	0	0	no change	N	N	N	N
0	1	0	0	1	*	*	*	*	*
0	1	0	1	0	no change	N	N	N	N
0	1	0	1	1	no change	N	N	Y	Y
0	1	1	0	0	no change	N	N	N	N
0	1	1	0	1	*	*	*	*	*
0	1	1	1	0	no change	Y	N	N	N
0	1	1	1	1	change	N	Y	Y	N
1	0	0	0	0	no change	N	N	N	N
1	0	0	0	1	*	*	*	*	*
1	0	0	1	0	no change	N	N	Y	N
1	0	0	1	1	no change	N	N	N	Y
1	0	1	0	0	no change	N	N	N	Y
1	0	1	0	1	*	*	*	*	*
1	0	1	1	0	change	Y	N	Y	Y
1	0	1	1	1	no change	N	Y	N	Y
1	1	0	0	0	no change	N	N	N	N
1	1	0	0	1	*	*	*	*	*
1	1	0	1	0	no change	N	N	Y	N
1	1	0	1	1	no change	N	N	Y	Y
1	1	1	0	0	no change	N	N	N	Y
1	1	1	0	1	*	*	*	*	*
1	1	1	1	0	change	Y	N	Y	Y
1	1	1	1	1	change	N	Y	Y	N

Y = yes, N = no, * = invalid start state

Fig. 2.11 OTF calculation table for the SN7473.

A word of warning, however: it is very easy to make mistakes in deciding whether each line results in fault-effect propagation or not. Propagation takes place if the presence of the faulty input causes the output to change when it should not have done so, or if the output does not change when it should have done so.

Both circumstances allow detection of the fault condition. Each line in Fig. 2.11 can be considered against these criteria by the simple expedient of altering the polarity of the defined value on each bona-fide input to the device and observing the resulting effect at the Q (and $\bar{Q}$) output.

For example, the J = 0, K = 1, T = 1, C = 1, Q = 1 line is evaluated separately for J s-a-1, K s-a-0, T s-a-0 and C s-a-0. In the absence of the fault, the Q output will change from 1 to 0. With J s-a-1, the Q output will still change and the fault will therefore not be detected. With K s-a-0, no change on Q will occur so the fault is detected, the same being true for the fault T s-a-0. Finally, the fault C s-a-0 will not be detected.

Note also that, as with the CTF calculations, the possibility of unstable states exists. In the case of the example flip-flop, 8 of the 32 possible start states, corresponding to C = 0, Q = 1, are invalid. The implication of this is that the total number of states is 24, not 32. In other words, the total number of sensitive or insensitive paths from a particular input is equal to the total number of valid states and not the apparent total number of states.

With this in mind, the various OTFs for this device can be calculated as follows:

Sensitive input	SP to Q output	IP to Q output	OTF
J	4	20	0.167
K	4	20	0.167
CLK	8	16	0.333
CLR	10	14	0.417

By symmetry, the OTF of the $\bar{Q}$ output is the same.

2.3.3 Computing input OY values

Returning to the way that OYs are related across devices, consider Fig. 2.12. If we let

OY(A-A) be the OY of node A at node A
    (defined as having the value 1)

OY(A-B) be the OY of node A at node B

OY(B-C) be the OY of node B at node C

OY(A-C) be the OY of node A at node C

then it must be true that:

OY(A-C) = OY(A-A) x OY(A-B) x OY(B-C)

That is, the transfer of OY values is multiplicative. Note that this property does not apply to CYs. CYs are absolute values fixed from the PIs of the circuit whereas OYs relate to a

particular source point and are refered to a particular path.   If
this property is to hold then we require that:

OY(A–B) = OY(A–A) x 'difficulty of transfer factor for
            device 1 to output B'

OY(B–C) = OY(B–B) x 'difficulty of transfer factor for
            device 2 to output C'

The 'difficulty of transfer factor' is the product:

OTF x g(CYs of supporting inputs)

where the function g is the 'average' of the supporting input's
CYs, calculated in the same way as the function f in the CY
calculation – expression 2.7.

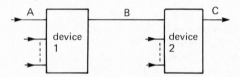

Fig. 2.12 Sensitivity transfer across two devices.

The generalised formula for OY transfer is, then:

OY(I–O) = OY(I–I) x OTF(I–O) x CY(av)

where

$$CY(av.) = \frac{1}{(i+j)} \; [\Sigma \, (CY: i \text{ async supporting inputs})$$

$$+ \left\{ (CY:clock) \times \Sigma \, (CY: j \text{ sync supporting inputs}) \right\}] \quad (2.11)$$

In the event that the node in question (node I) is a clock node
for the device concerned, then the clock CY in equation 2.11 is
set to 1.
    The earlier discussion on OY transfer would suggest   that   the
procedure to be used in the calculation is as follows.   Start at a
device output node, where the OY is 1 and transfer this value to
the circuit's primary outputs to obtain the values for the OYs of
the device at the primary outputs. This process should then be
repeated for each circuit node.
    An objection to this technique is that it is time consuming,
since the calculation must be repeated as many times as there are
circuit nodes. It will also be complicated by factors such as
reconvergent fanout and global feedback, both of which will cause
simultaneous equations to be produced.
    Fortunately, a simpler technique can be used which is based on

the multiplicative property of OY transfer.
Returning to Fig. 2.12, we have:

$$OY(A\text{--}C) = OY(A\text{--}A) \times OY(A\text{--}B) \times OY(B\text{--}C)$$

which is mathematically equivalent to:

$$OY(A\text{--}C) = OY(C\text{--}C) \times OY(B\text{--}C) \times OY(A\text{--}B)$$

since both $OY(A\text{--}A)$ and $OY(C\text{--}C)$ are 1. This suggests an alternative method of calculation – namely, to start at the circuit's primary outputs and work back through the circuit calculating each nodal OY value as the node is encountered. This calculation requires a single pass per primary output, and is much quicker.

The multiplicative property of OYs can be more easily understood by considering the sensitive inputs and outputs of devices on the sensitive path to be connected by a relay contact. If the contact is closed, then the path propagates; if it is open then the path is blocked. The chance of the operation of each contact is determined by the CYs of the supporting inputs to the device concerned and its transfer factor. Thus, the ease of completing the whole path is determined by multiplying the control factors for each device together. This result is independent of the ordering of the terms involved in the multiplication.

2.3.4 Output fanout

The presence of fanout may make a device output observable at several circuit outputs, as shown in Fig. 2.13. Here, the node X can be observed both at PO1 and at PO2 and we can calculate both $OY(X\text{--}PO1)$ and $OY(X\text{--}PO2)$. How are these two OYs to be combined to form a composite total OY for the node? The answer is provided by comparing the situation with that of parallel-path redundancy in reliability calculation. For this situation, success is guaranteed if at least one path is operating, i.e:

$$R(S) = 1 - Q(S) = 1 - \Pi \; [Q(\text{each path})]$$

$$= 1 - \Pi \; [1 - R(\text{each path})]$$

where R is the reliability of a process and Q is its unreliability.

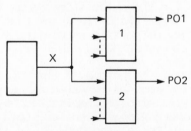

Fig. 2.13 Fanout to several primary outputs.

Translating this into OY terminology, we are saying that, provided at least one of the path options can be sensitised, the node can be successfully observed at a primary output. If more than one path can be sensitised, then:

OY(composite) = 1 − Π [1 − OY(each PO)]

which, for the example in Fig. 2.13, gives:

OY(X−composite) = 1 − [1 − OY(X−PO1)][1−OY(X−PO2)]

2.3.5 Path reconvergence

There exists the possibility that fanout paths will reconverge within a circuit. There are two cases to consider: reconvergence via unequal or equal path lengths.

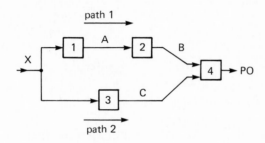

Fig. 2.14   Reconvergence via unequal path lengths.

For unequal path lengths (Fig. 2.14), the strategy is to select the OY(X−PO) corresponding to the shortest path from X to PO, i.e. path 2 in Fig. 2.14. The assumption here is that, in practice, the shortest path would be selected for sensitisation and that other paths would be blocked to avoid the possibility of reconvergence.

For this example, the OY calculations would proceed in the following manner:

Stage	Path 1	Path 2
1:compute	OY(B−PO)	OY(C−PO)
2:compute	OY(A−PO)	OY(X−PO)
3:no further action		

At stage 3, a non−zero value has been calculated for the observability of node X via path 2, so there is no need to complete the calculation of the OY via path 1.

For reconvergence via equal path lengths (Fig. 2.15), the strategy is to compute OY(X−PO) for both paths and to retain the higher value. Again, the assumption is made that all but one of the possible paths from X to PO can be blocked.

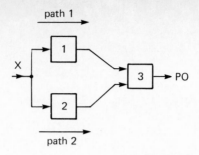

Fig. 2.15 Reconvergence via equal path lengths.

2.3.6 Feedback paths

For feedback paths, the strategy is derived by considering the situation to be the same as that of reconvergence with unequal path lengths. In Fig. 2.16 the sequence of computations would then be:

Stage	OY value computed	Comment
1	OY(C–PO)	
2	OY(B–PO)	
3	OY(A–PO)	OY(C–PO) is not recalculated: answer exists

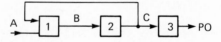

Fig. 2.16 Feedback.

A possible objection to this strategy is that, in calculating OY(A–PO), we have assumed controllability on the other input to device 1. With the structure shown in Fig. 2.16 this assumption may be invalid depending on the way that the feedback line is able to control the output of device 1. The validity of the assumption increases if there are other devices in the feedback path because the possibility of being able to block the propagation of fault–effect information around the feedback path is increased. In effect, this is a compromise between the desires to be accurate and to be practical.

2.4 CONCEPTS OF TESTABILITY

Let us summarise what has been achieved so far.

First a method for calculating the controllability of every node in a circuit has been developed. This produces a figure of merit, between 0 and 1, which reflects the ease or otherwise with which the logic value of the node can be controlled from the circuit's primary inputs.

Second, a method for calculating observability values for every circuit node has been developed. The figure of merit, again varying between 0 and 1, reflects the ease or otherwise with which the change in a node's logic value caused by a fault can be observed at the circuit's primary outputs.

Both the controllability and observability measures relate to nodes in the circuit, rather than to devices, although the devices are instrumental in dictating the values produced for each node. Testability must, therefore, also be related to circuit nodes and should reflect the ease, or otherwise, of being able to generate tests for fault conditions present on a circuit node (caused by a nodal or device failure mechanism). This means that testability must be a composite function of both controllability and observability.

2.4.1 TY as a function of CY and OY

A simple measure of testability (TY) can be obtained from the product of CY and OY values for a node. The relationship employed in CAMELOT is:

$$TY \text{ node} = CY \text{ node} \times OY \text{ node} \qquad (2.12)$$

which satisfies the requirements that:

(a) TY = 0, if either CY or OY is 0;
(b) TY = 1 if both CY and OY are 1;
(c) 0 < TY < 1 for 0 < CY < 1 and 0 < OY < 1.

It also satisfies the intuitive feeling that if a node is, say, 50% controllable (CY=0.50) and 50% observable (OY=0.50), then its testability is only 25% (TY=0.25) rather than 50%. This is because controllability and observability are independent quantities; if it is '50% more difficult' to control a node and '50% more difficult' to observe it, then the resultant testability should be somewhat less than 50%.

What, then, is the overall testability value for the circuit? This should be a measure of the average difficulty of producing a test for a node in the circuit and, therefore, the value used is the arithmetic mean of the individual nodal TYs, i.e.:

$$TY(\text{circuit}) = \frac{\Sigma\,(TY\text{: nodes})}{\text{No. of nodes}} \qquad (2.13)$$

2.4.2 Computation procedure

Briefly, then, the procedure followed in CAMELOT for calculating testability values is as follows.

STEP 1 Prepare, read in and check a description of the circuit connectivity.

STEP 2 Calculate nodal CY values, starting at the circuit's primary inputs and progressing through the circuit. This requires both the connectivity description and a library containing device CTF values.

STEP 3 Calculate nodal OY values, starting at the circuit's primary outputs and working back towards its primary inputs. Here, the library is interrogated to obtain OTF values for the devices in the circuit.

STEP 4 Calculate nodal TY values from the nodal CY and OY values.

STEP 5 Calculate and present the circuit's average testability and interpret the results.

2.5   PRACTICAL USE OF CAMELOT

Testability measures, such as CAMELOT, can be used both by logic designers and by test programmers. Prediction of the controllability, observability and testability profiles allows the user to observe and learn about the testability characteristics of the design. Specifically, a designer is able to identify areas of low controllability and low observability and to assess the effect of design changes in terms of testability gain. Test programmers can also benefit from a knowledge of the TY profile in that it can form the basis for determination of the testing strategy.

This section explores these ideas further and comments on other practicalities of using CAMELOT.

The CAMELOT results presented in the following subsections have been produced by the Cirrus Computers' implementation of the CAMELOT theory. More detail of this program, including the methods for handling bidirectional nodes, tristate nodes, wired nodes and bus-structures, can be found in reference 2.2, listed in the bibliography at the end of the chapter. One specific point to note however is that the CY, OY and TY values presented to the user and listed in these sections are the eighth root of the calculated values. This is to confine the scale of the absolute figures and to make the results easier to interpret.

2.5.1   Quantifying design changes

To illustrate how CAMELOT can be used to quantify the effect of design changes, consider the divide-by-10 ripple counter shown in Fig. 2.17. An initial CAMELOT analysis of the circuit yields the CY, OY and TY values for all the circuit nodes, listed in Fig. 2.18.

Figures 2.19-2.21 present these results in histogram format and, before continuing with this particular example, we should make some general observations about the interpretation of the figures.

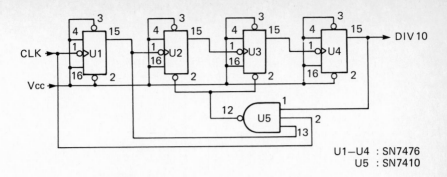

Fig. 2.17 Divide-by-10 ripple counter.

Node	Testability	Controllability	Observability
U5.12	0.360	0.607	0.593
U2.15	0.370	0.675	0.548
U1.15	0.373	0.795	0.469
CLK	0.432	1.000	0.432
U3.15	0.480	0.641	0.750
Vcc	0.660	0.841	0.785
DIV10	0.692	0.692	1.000
Mean values	0.481	0.750	0.654

Fig. 2.18 TY values for basic counter.

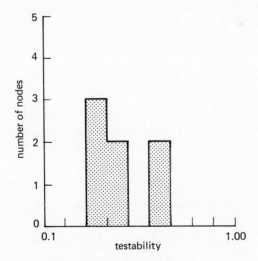

Total number of nodes : 7
Mean testability value : 0.481
Worst rated node   : U5.12 (TY = 0.360)

Fig. 2.19 Distribution of testability values.

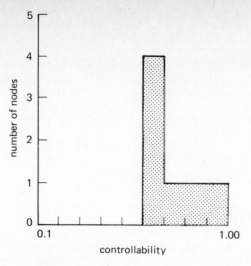

Total number of nodes   : 7
Mean controllabilty value : 0.750
Worst rated node   : U5.12 (CY = 0.607)

Fig. 2.20 Distribution of controllability values.

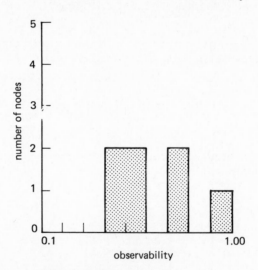

Total number of nodes   : 7
Mean observabilty value : 0.654
Worst rated node   : CLK (OY = 0.432)

Fig. 2.21 Distribution of observability values.

   The values calculated are primarily intended as a means of
comparing testabilities within a particular circuit; they do not
provide an accurate means of comparing the testabilities of
different circuits. Note that the CY values for a circuit affect
the OY values calculated. Changes to a circuit that alter its

controllability properties should therefore be made before any changes based on its observability properties are considered.

Poor controllability can be remedied by allowing more direct control either over the affected area of the circuit, or over an area preceding it (i.e. in the signal flow path between the primary inputs and the affected area). Poor observability can be remedied by allowing easier access either to the affected area, or to an area between it and the circuit's primary outputs (e.g. by adding extra monitor points).

In reality, the ripple counter example is easily tested as it stands. Nevertheless, to illustrate how CAMELOT can be used to assess the effect of design changes, we will consider ways of improving the testability of the circuit.

Figure 2.18 identifies those nodes in the circuit where, because of poor controllability and/or poor observability, testability is likely to be poor. Note that OY values are consistently lower than CY values. This is intuitively correct; the observation task requires not only the identification of suitable sensitive paths but also the creation of logic states in the circuit that will support them

From Fig. 2.18, it can be seen that the least testable node is U5.12. This node exhibits problems in both control and observation. To control the node, only one state of the counter will cause the node to go low compared with nine states to send it high. This places restrictions on testing the stuck-at-one state. To observe the node, the effect of presetting U2 and U3 must be propagated through the rest of the counter to the output DIV10. These problems also account for the next least testable node U2.15 which can be controlled from U5.12 and only observed via the counter output.

If Design-For-Testability guidelines (such as those described in Chapter 5) are applied to the circuit, then their effect on circuit testability can be evaluated. Two recommendations for this circuit would be:

(1) provide links to break and therefore control and observe any global feedback paths;
(2) increase access to the inputs or outputs of internal stored-state devices.

To implement recommendation (1), the connection from U5.12 to the preset inputs of U2 and U3 can be broken and the two ends connected external to the circuit, i.e. by routing the two ends to unused device pins, if the circuit is realised on a single integrated circuit, or to unused edge connector fingers if the circuit is realised on a PCB. This modification increases both the controllability of the counter chain and the observability of the logic feeding it. In normal operation, the connection would be completed by linking the appropriate chip pins or back-plane pins.

To implement recommendation (2), an extra monitor point connected to U2.15 would suffice. The revised circuit is shown in Fig. 2.22. The effect of these changes on the testability profile of the circuit is shown by the improved testability figures, listed in Fig. 2.23 and shown in histogram form in Fig. 2.24.   In this case, the mean testability has increased from 0.481 to 0.734, a gain of 52% for only a modest revision to the basic design.

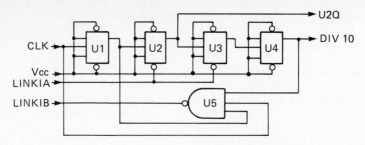

Fig. 2.22   Revised divide–by–10 ripple counter.

Node	Testability	Controllability	Observability
LINK1B	0.611	0.611	1.000
U3.15	0.625	0.783	0.799
U1.15	0.708	0.795	0.890
DIV10	0.724	0.724	1.000
Vcc	0.773	0.841	0.919
CLK	0.784	1.000	0.784
U2Q	0.785	0.785	1.000
LINK1A	0.860	1.000	0.860
Mean values	0.734	0.817	0.906

Fig. 2.23   CAMELOT values for modified circuit.

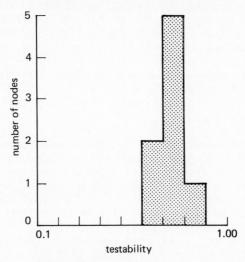

Total number of nodes   : 8
Mean testability value   : 0.734
Worst rated node   : LINKIB (TY = 0.611)

Fig. 2.24   Revised distribution of testability values.

2.5.2 Automatic selection of test points

The CY, OY and TY rating figures produced by CAMELOT can be used to position test points automatically around a circuit so that the optimum testability improvement results. Here, a test point is defined to be a node to which access is available from a test system for the sole purpose of observing the logical state of the node. The test point is not one of the normal operating interconnections of the circuit; it is used only during the processes of testing and diagnosis.

This section comments briefly on the theory behind the test-point placement algorithm and illustrates its outcome on a circuit more complex than that used in the previous section. More detail of the algorithm can be found in Section 9 of reference 2.2.

An exhaustive process for identifying the best set of N nodes of a circuit at which to place test points would be to measure the effect on the circuit's testability of placing test points at each possible node and to select that set which produces the greatest improvement. Whilst this technique would be guaranteed to produce the optimum set of nodes, it would require an excessive computer run time and is therefore impractical. Any other technique is not guaranteed to produce the optimum set of nodes, but may produce a satisfactory set within an acceptable run time.

Thus, an alternative is to develop a repetitive algorithm that locates a small number of test point sites on each pass so as to produce the best testability improvement at that stage.

The algorithm starts by examining the ordered list of OY values so as to locate the least-observable node in the circuit. Test access points improve circuit testability by producing improvements in the OY ratings of internal circuit nodes. They have no effect on CY values. Thus the least-observable node is a good place to begin the search for possible test point locations.

In practice, it has been found that the least observable node is not the most suitable place to locate the test point. This is more likely to lie on a node some distance forward along the path of signal flow through the circuit. In the context of the CAMELOT theory, this will be a node where, because of poor supporting CY or OTF or both, sensitivity transfer to the next node is likely to be difficult. Alternatively, in circuits without significant contrast in supporting CY or OTF values, the most suitable node will be that node which produces the best overall improvement in the circuit as a whole. In either case, it is necessary to treat all nodes on the shortest forward path from the least-observable node as possibilities in the search for a test point.

Having located all the possible nodes at which a test point could be placed, a means must be developed to select between them based on the success with which they can improve the circuit's testability profile. As a result, a single node will in most cases be identified as the best location for a test point. However, in cases where some degree of circuit symmetry exists, it is possible that several nodes will tie for top place and on these occasions more than one node may be assigned as a test point.

At the close of the pass of the algorithm, the selected nodes are inserted into the circuit description as test points and the circuit's testability profile is recalculated. The algorithm can then be repeated until the desired number of test points has been inserted.

Briefly, then, the algorithm consists of three stages.

1. Identify a set of candidates, and rate them according to the effect they would have on the circuit's testability profile if they were to be made test points.

2. On the basis of the candidate ratings, select the most suitable nodes to be assigned as test points, having due regard to the number of test points that remain to be assigned.

3. Insert the selected test points into the circuit description and calculate the new testability profile.

The following example illustrates the application of the algorithm to the sequential controller circuit shown in Fig. 2.25. Figure 2.26 lists the full set of nodal TY, CY and OY values for this circuit and Fig. 2.27 shows the testability profile.

A CAMELOT automatic test-point selection run on this circuit produced the following recommendations for test points, listed in order of decreasing importance:

Test point		Mean nodal testability	Cumulative % increase in testability
None placed		0.518	0.00
U11.10	(TP1)	0.544	5.02
U7.8	(TP2)	0.565	8.88
U12.6	(TP3)	0.582	11.89
U12.8	(TP4)	0.595	14.12
U8.6		0.602	15.30
U7.6		0.612	16.97
U9.8		0.618	17.95
U10.3		0.622	18.60

Of these points, the first four are significant in terms of their contribution to the cumulative percentage increase in testability. These points are shown as test points TP1–TP4 in Fig. 2.25 and, as can be seen, are reasonably dispersed around the circuit and are all placed on significant feedback paths. The effect of these test points on the testability profile of the circuits is shown in Fig. 2.28.

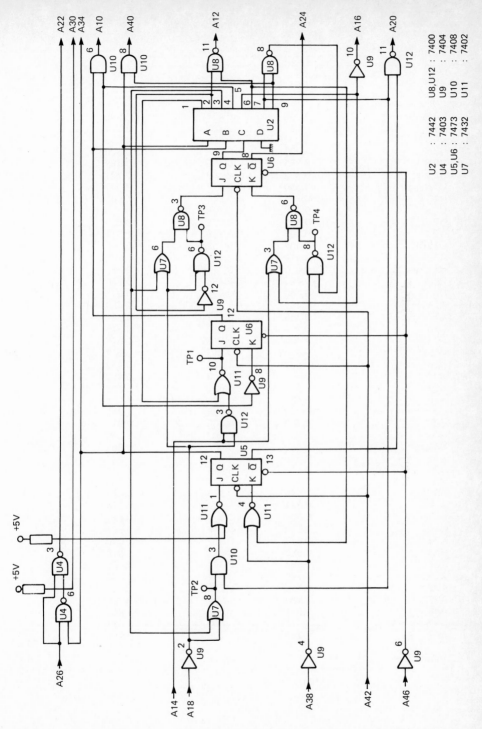

33

Fig. 2.25 Sequential controller circuit (courtesy GenRad Inc.)

34

Node	Testability	Controllability	Observability
U8.8	0.245	0.411	0.595
U2.1	0.281	0.488	0.574
U9.12	0.291	0.488	0.595
U2.3	0.303	0.488	0.620
U9.8	0.333	0.488	0.681
U2.2	0.352	0.488	0.721
U7.8	0.360	0.717	0.501
U12.3	0.371	0.841	0.441
U2.9	0.376	0.488	0.770
U2.7	0.395	0.488	0.809
U2.4	0.396	0.488	0.811
U10.3	0.404	0.532	0.759
A12	0.411	0.411	1.000
U11.10	0.413	0.611	0.675
U2.6	0.425	0.488	0.870
U5.13	0.443	0.691	0.641
A14	0.459	1.000	0.459
U12.8	0.460	0.712	0.647
U7.3	0.462	0.717	0.645
U7.6	0.464	0.717	0.647
U12.6	0.464	0.717	0.647
A18	0.469	1.000	0.469
U9.2	0.469	1.000	0.469
U8.6	0.488	0.601	0.813
A16	0.488	0.488	1.000
U2.5	0.488	0.488	1.000
U8.3	0.490	0.603	0.813
A10	0.503	0.503	1.000
U11.1	0.509	0.565	0.901
A40	0.516	0.516	1.000
A20	0.516	0.516	1.000
U6.9	0.574	0.677	0.848
U6.12	0.579	0.668	0.867
A38	0.624	1.000	0.624
U9.4	0.624	1.000	0.624
U11.4	0.639	0.717	0.891
A24	0.677	0.677	1.000
GND	0.688	0.841	0.818
A34	0.691	0.691	1.000
A30	0.744	0.744	1.000
A22	0.756	0.756	1.000
A26	0.885	1.000	0.885
A46	0.911	1.000	0.911
U9.6	0.911	1.000	0.911
A42	0.957	1.000	0.957
Mean values	0.518	0.678	0.776

Fig. 2.26 TY, CY and OY values for sequential controller.

35

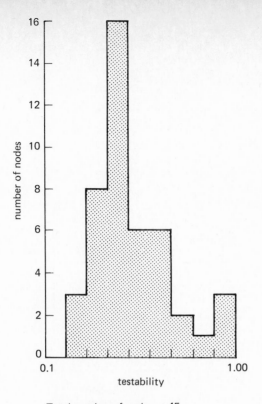

Total number of nodes : 45
Mean testability value : 0.518
Worst rated node : U8.8 (TY = 0.245)

Fig. 2.27 Testability profile for sequential controller.

2.5.3 Planning a manual test-generation strategy

Planning a manual test-generation strategy for an unstructured design (at chip or board level) is very much a neglected topic. The key question is usually - where to start? In answering the question, there is generally a number of options.

Option 1 Start with the hardest node to control.
Option 2 Start with the hardest node to observe.
Option 3 Start with the easiest node to control and observe, i.e. to test.

The justification for either option 1 or option 2 is that in tackling, and hopefully solving, a difficult problem first, many of the easier faults will automatically be covered in the process. The justifiction for option 3 is that solving an easy problem first provides a gentler introduction to the intricacies of the circuit and, since the problem is easily solved, boosts the morale of the test programmer!

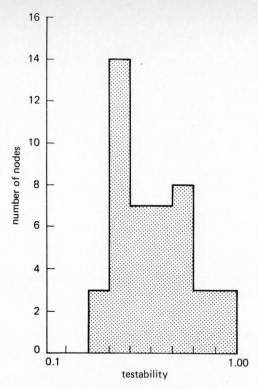

Fig. 2.28 Testability profile for modified sequential controller.

Total number of nodes   : 45
Mean testability value   : 0.595
Worst rated nodes   : U9.8 (TY = 0.371)
                           U8.8 (TY = 0.383)

Both approaches provide a useful way of structuring the
approach to the test—generation process. Too often however, the
strategy for testing a circuit is developed in a relatively ad hoc
way. Nodal testability figures provide a way of planning the
strategy in a more scientific way.

To illustrate this, consider again the sequential controller
circuit (Fig. 2.25) together with the nodal TY, CY and OY values
computed by CAMELOT (Fig. 2.26). The nodes with the lowest
controllability in this circuit are nodes A12 and U8.8, both of
which have a CY value of 0.411. A12 is the NAND result of outputs
U2.2 and U2.6 of the decoder U2. U8.8 is the NAND result of
outputs U2.7 and U2.9 of U2. A12 is directly visible as a primary
output whereas U8.8 is fed back into the decoder control
circuitry.

This information alone is sufficient to suggest that the
decoder U2 is a significant component of the circuit and perhaps
we should concentrate the test—generation strategy on this device
(a deduction which is obvious for this particular circuit but
which may not be so obvious in a substantially larger circuit).

Now consider the observability figures for the circuit. The node with the lowest observability is U12.3 with an OY of 0.441. This node is logically (and topologically) in the centre of the circuit. Its value can be observed most directly at the A10 primary output but also at other POs by virtue of the connection from U6.12 to U2 and hence onto the output side of the decoder U2. We may decide therefore to start the test generation procedure by deriving tests for U12.3 both s-a-1 and s-a-0 on the grounds that in creating the sensitive paths, we will have tackled and solved one of the most complex test-generation problems in the circuit.

Finally, let us move to the other end of the spectrum - the easiest nodes to test. Here, our interest is to determine the node which is both easy to control and to observe, i.e. the node with the highest testability rating. In this case, the primary input A42 comes out as the easiest testable node (TY = 0.957) with the primary input A46 a close second (TY = 0.911). This is hardly surprising since A42 is the main clock input and A46 the general reset. The result is not too helpful therefore. We expect clock and reset inputs to be easy to test simply because they have a major effect on the behaviour of the circuit. Nevertheless, it is encouraging when the theoretical figures produce a result which, intuitively, we know is correct.

In summary therefore, the analysis of the testability rating figures would suggest that the best approach to testing the circuit is to concentrate on the U2 decoder device. Its output affects almost every other device in the circuit and learning to control and observe the decoder will inevitably allow control and observation of the rest of the circuit. In practice, this is a reasonable strategy for this circuit.

2.5.4 Planning an automatic test-generation strategy

It is only very recently that research workers have begun to study the use of nodal testability figures in terms of their ability to steer the progress of automatic test-generation algorithms such as the D-algorithm or the critical-path algorithm. The practical implementation of such algorithms has been complicated by the problems of choice - choice of target fault; choice of fault-effect transmission path; choice of device inputs and input values to sustain or block a particular sensitive path. Traditionally, choosing from amongst a number of options to satisfy a particular requirement has either been random or has been governed by some other local or global selection criterion. An alternative approach is to base the selection on the nodal OY, CY and TY values. For example, the choice of the starting target fault could be based on the highest testable node. Similarly, forward path choice decisions could be based on a more detailed analysis of routes of highest observability (to maximise the chance of successful propagation of the fault-effect to an observable primary output). In addition, nodal controllability values can be used to indicate the 'best' strategy for satisfying a particular nodal fixed-value requirement, e.g. during the consistency phase of the D-algorithm or during the backtrace operation of the PODEM algorithm. The latter application is explored in more detail in Chapter 4. In general however, a detailed discussion of the use of testability ratings to control the application of a test-generation algorithm is outside the scope of this book. Interested readers are referred to references 2.4 and 2.32.

2.6 COMMENT ON OTHER TESTABILITY MEASURES

This chapter has described, in some detail, the concepts and application of the CAMELOT program. Before we leave the subject of testability measures, it is appropriate to mention other work in the area.

2.6.1 TMEAS (Stephenson and Grason, references 2.5, 2.6, 2.7)

As in CAMELOT, CY and OY values in TMEAS are normalised in the range 0 (bad) to 1 (good). In order to calculate the circuit CY values, the inputs are first assigned CYs of 1. CYs at the remaining circuit nodes are then related to those at the inputs by a set of simultaneous equations. These equations are obtained by using CTFs to relate the CYs on a component's inputs to those on its outputs. The CTFs vary according to the logical function of the component and are a function of the uniformity of the input-output map. The CTF of a component with a uniform input-output mapping is 1, decreasing to 0 as the degree of uniformity decreases.

In contrast to CAMELOT, TMEAS calculates OY values for the circuit by first assuming that the primary outputs have OY = 1. OYs for the other circuit nodes are then related to the output OYs by using OTFs. These OTFs measure the probability that a faulty value at any of the component's inputs will propagate to its outputs and, again, values vary between 0 and 1 according to the logical function performed. The OY calculation in TMEAS is completely independent of the CY values obtained previously. In both CY and OY cases, special rules are developed for dealing with stored-state components, clock lines and fanout. Briefly, these are as follows. Stored-state devices are modelled by the addition of feedback paths around the device which represent the dependence of the device's next state on its current stored state. The presence of master clock lines is ignored. And, for the controllability calculation, the CY given to fanout branches is reduced from that of the trunk to model the fact that the branches cannot be set to differing values as might be desirable during path sensitisation. These rules differ from those used in CAMELOT. TY values are obtained from the CY and OY figures by taking their geometric mean.

A limitation of TMEAS, which is also present in CAMELOT, is that the formulae given for the calculation of CTF and OTF values from the device function tend to produce unrealistically low values for the simpler circuit elements (AND, OR, etc). This is not serious, though, as the values can always be increased to some arbitrary level if problems are caused. A further difficulty, again common to TMEAS and CAMELOT, is that of dealing with circuit nodes that are permanently tied to one or other logic level. This arises because the calculation of a single CY value for each node tends to hide the relative difficulties of producing the two possible logic levels. The following three measures offer solutions to the problem.

2.6.2 TEST/80 (Breuer and Friedman, references 2.8, 2.9)

TEST/80 is a proposed automatic test-generation program for digital circuits. Like the D-algorithm, it attempts to set up paths through the circuit by means of which a fault at an internal node may be observed at the circuit's outputs. This is done in two phases; setting up a path from the node to an output and establishing a set of input conditions to support the path. In contrast to the D-algorithm, TEST/80 employs a process, called 'cost analysis', which enables it to identify the optimum path to follow at any stage. This cost analysis is, in effect, an evaluation of the controllability and observability properties of the circuit.

Three cost values are calculated for each circuit node. These are: cA, the cost of setting node A to a 1; c$\bar{\text{A}}$, the cost of setting the node to a 0; and dA, the cost of driving a D on node A to an output of the circuit. As the various problems become more difficult to solve, so the costs rise.

In common with the SCOAP and TESTSCREEN measures described below, controllability values are calculated separately for each of the logic levels on a node – in this case cA and c$\bar{\text{A}}$. The values of these costs at the device outputs are produced, with a knowledge of the device's function, by summing three factors:

(a) a function of the input controllability costs;
(b) a measure of the adverse effects that fanout from the device output might have;
(c) a factor associated with the device type.

By performing this calculation for each device in the circuit, a set of equations may be produced from which cA and c$\bar{\text{A}}$ values may be determined for all circuit nodes.

The third cost, dA, is a measure of the observability of the node. The value of this cost at the input to a device is again calculated by summing three factors.

(a) A measure of the probability of being able to propagate a D from the input to the output. This is a function of the cA and c$\bar{\text{A}}$ values of the other device inputs.
(b) The cost of propagating a D from the device output to the circuit output. Where several alternative paths exist the lowest cost is taken.
(c) A measure of the number of clock cycles that will be taken to transfer the D at the node to the circuit output.

It can be seen that the cost dA is zero if A is a circuit output. Note that the principle of employing controllabilities in the calculation of observability is as used in CAMELOT.

This measure produces testability parameters whose magnitudes decrease as the respective tasks become easier to perform. Thus, to maximise testability the costs must be minimised. On the other hand, the ability to calculate controllabilities separately for the two logic levels increases the accuracy with which the test generation process may be modelled and resolves the problems posed by permanently tied nodes.

2.6.3 SCOAP (Goldstein, references 2.10, 2.11)
COMET (Berg and Hess, references 2.12, 2.13)

SCOAP has evolved from TEST/80. The circuit is characterised by assigning six values to each node: combinational 0 and 1 controllabilities, sequential 0 and 1 controllabilities and combinational and sequential observabilities. These are defined as follows.

The combinational controllabilities, CC0 and CC1, are defined as the minimum number of nodes that must be set in order to produce either a 0 or a 1 on the node in question. Sequential controllabilities SC0 and SC1, are, similarly, the number of 'sequential nodes' that must be assigned. A sequential node is a circuit node held for one time period. Thus a node which has to be held high for six time slots in order to propagate a desired logic level to a given node counts as six sequential nodes. CC0 and CC1 are defined to be 1 at circuit inputs, whilst SC0 and SC1 are defined to be zero.

The combinational and sequential observabilities, CO and SO are defined as the minimum number of nodes or sequential nodes, respectively, which must be set for a fault to propagate from its source to a circuit output. Observabilities are defined to be zero at circuit outputs.

As in the TEST/80 measure, controllabilities and observabilities are related across devices by their logical function. Also, their values are inversely related to the circuit's properties.

A modified form of the SCOAP algorithm to suit a gate-array design environment has been described by Berg and Hess (references 2.12 and 2.13). Their system, referred to as COMET (Controllability and Observability MEasure for Test), includes a further testability parameter called predictability. Predictability is a derivative of controllability and provides a measure for the ease or otherwise of being able to initialise a circuit into a known start state.

Other modifications to the basic SCOAP algorithm include better handling of power and ground lines, tied nodes, bidirectional devices, and macro-level cells.

2.6.4 TESTSCREEN (Kovijanic, references 2.14, 2.15, 2.16, 2.17)

One deficiency of the SCOAP measure is that, for a string of n inverters, the output controllability is $(n - 1)$. Conceptually, however, it is no more difficult to control the output of n inverters than the output of one. To resolve this difficulty, observability definitions in TESTSCREEN are modified from SCOAP's definitions to be in terms of the number of circuit inputs that must be set to achieve the desired result. Thus, for example, combinational controllability is defined as the number of circuit inputs that must be set to produce the desired logic level on the node. Hence, for the inverter string, the output combinational controllabilities are now 1.

In TESTSCREEN testability is defined to be a weighted function of the six controllability and observability factors, the network size and the number of input/output pins.

2.6.5 VICTOR (Ratiu, reference 2.18)

The VICTOR testability analysis program has been developed by Ratiu with the aim of identifying redundant nodes in a circuit prior to the test-generation process. A circuit node is logically redundant if all the output values of a circuit are independent of the binary value on the node for all input combinations or state sequences. In practical terms, this means that certain faults on a logically-redundant node cannot be detected. Whether such faults need to be detected is an interesting question explored further in Section 5.1 of Chapter 5 (Guideline 2). The issue with testability analysis is whether the potential redundancy should be identified before attempting test generation. In this way, time-consuming and ultimately fruitless computation will be avoided.

VICTOR, standing for VLSI Identifier of Controllability, Testability, Observability and Redundancy, analyses the fanout and reconvergence structure of the combinational section of a scan-designed circuit (described in the next chapter). The program provides a warning of the existence of a potentially-redundant node based on a potential conflict of desired logic values on the node. The user can then chose either to modify the circuit or to instruct the test generator to ignore faults on the node.

2.6.6 'Entropy' measure (Dussault, reference 2.19)

Dussault's measure is based on information theory and, as such, bears little relationship to the measures discussed previously. Whilst being of interest from a theoretical viewpoint, the measure is described in terms of combinational circuits only and it would appear that it would be difficult for it to be applied to sequential devices. This would considerably restrict its application.

Dussault defines controllability and observability as entropy-like variables. At circuit inputs it is known exactly what logic signals are being applied and, therefore, the controllability is high. As signals progress through the circuit, so the knowledge of the node's signal value decreases, thus decreasing controllability. Similarly, observability is high at circuit outputs and decreases towards the circuit's inputs. These definitions broadly agree with those used by CAMELOT and TMEAS.

2.6.7 An overall comparison

Each of the measures described above differs from CAMELOT in a number of respects, as discussed.

Whilst some of these differences will tend to increase the accuracy of the testability predictions, they do so at the expense of increased theoretical complexity. An example is the calculation of separate testability parameters for the two logic states, as in TEST/80, SCOAP and TESTSCREEN, which requires a knowledge of a device's logical behaviour during the calculation.

Other features tend to reduce the accuracy of the predictions, an example here being the independence of the observability calculations in TMEAS from the previously calculated controllability values.

Clearly, the balance between complexity and accuracy must be selected according to the judgement of the developers and the use to which the testability predictions are to be put. CAMELOT represents one such balance, considered satisfactory for use in a practical design environment. The other measures represent a different balance.

2.7 CONCLUDING REMARKS ON TESTABILITY MEASURES

The chapter has discussed a testability measure based on the concepts and quantification of nodal controllability and observability values. Of necessity, such a measure can only produce coarse results since, in reality, the only real measure of testability is the cost of producing an adequate set of tests for the circuit. This observation raises a basic question – do measures such as CAMELOT have any real value? The author believes the answer to this question is 'yes', provided the user understands the limitations of the theory. This view is supported by the growing number of companies now using testability analysis programs, of which CAMELOT is an example, as part of their design process. On their own, the CY, OY and TY nodal values obtained by CAMELOT have very little meaning. Comparatively however, the results are meaningful. A node with a CY value of 0.4 will not be so easy to control (to one level at least) as another in the same circuit, with a CY value of 0.8. Similarly, if changes are made to the circuit design which result in an increase in overall testability, from 0.5 to 0.7 say, then the percentage increase (40%) is a useful way of quantifying the value of the change, at least from a testability point of view. This valuation can be judged against other cost criteria such as the cost of extra components or primary inputs and outputs.

Another significant advantage of testability measurement tools, such as CAMELOT, is their educational use. An analysis of CAMELOT-generated results focusses the attention of the designer onto those sections of the circuit which are of low controllability and low observability. In this way, the designer becomes conscious of areas of the circuit which are potential problems for the test-programmer and can decide to reduce the severity of the problem by modifying the design before committing to silicon or printed-circuit board. In carrying out this process, the designer learns about the more practical aspects of testability such that, in the future, his or her awareness of the meaning of design for testability is increased. This surely is one of the ultimate objectives of any design-for-test aid.

BIBLIOGRAPHY

CAMELOT

2.1 Bennetts R.G., Maunder C.M., and Robinson G.D., 1981, "CAMELOT: a computer-aided measure for logic testability", IEE Proc., Vol. 128, Part E, No .5, pp 177-189.

2.2 CAMELOT User Manual, 1983, available from Cirrus Computers Ltd. (29/30 High Street, Fareham, PO16 7AD, UK).

2.3 Ibbotson J.B., 1983, "Testability measurement: use of CAMELOT", Test!, Vol. 5, No. 2, March, pp 29-34.

2.4 Maunder C.M., 1983, "Using testability measures in design and test", Proc. Network ATE Conf., Brighton, Dec.

TMEAS

2.5 Stephenson J.E., 1974, "A testability measure for register transfer level digital circuits", Ph.D. dissertation, Carnegie-Mellon University.

2.6 Stephenson J.E., and Grason J., 1976, "A testability measure for register transfer level digital circuits", Proc. 6th IEEE Fault Tolerant Computing Symposium, pp 101-107.

2.7 Grason J., 1979, "TMEAS - a testability measurement program", Proc. 16th IEEE Design Automation Conf., pp 156-161.

TEST/80

2.8 Breuer M.A., 1978, "New concepts in automated testing of digital circuits", Proc. EEC Symposium on CAD of Digital Electronic Circuits and Systems, Brussels, pp 69-92.

2.9 Breuer, M.A., and Friedman A.D., 1979, "TEST/80 - a proposal for an advanced automatic test generation system", Proc. IEEE Autotestcon, pp 205-312.

SCOAP

2.10 Goldstein L.H., 1979, "Controllability/observability analysis for digital circuits", IEEE Trans. Circuits and Systems, CAS-26, No.9, pp 685-693.

2.11 Goldstein L.M., and Thigpen E.L., 1980, "SCOAP: Sandia Controllability/Observability Analysis Program", Proc. IEEE 17th Design Automation Conf., pp 190–196.

COMET

2.12 Berg W.C., and Hess R.D., 1982, "COMET: a testability analysis and design modification package", Proc. IEEE Test Conf., Paper 13.1, pp 364–378.

2.13 Hess R.D., 1982, "Testability analysis: an alternative to structured design for testability", VLSI Design, March/April, pp 22–27.

TESTSCREEN

2.14 Kovijanic P.G., 1979, "Testability analysis", Proc. IEEE Semiconductor Test Conf., pp 310–316.

2.15 Kovijanic P.G., 1979, "Computer aided testability analysis", Proc. IEEE Autotestcon, pp 292–294.

2.16 Kovijanic P.G., 1981, "Single testability figure of merit", Proc. IEEE Test Conf., Paper 18.1, pp 521–529.

2.17 Dunning B., and Kovijanic P., 1981, "Demonstration of a figure of merit for inherent testability", Proc. IEEE Autotestcon, pp 515–520.

VICTOR

2.18 Ratiu I.M., et al. 1982, "VICTOR ; a fast VLSI testability analysis program", Proc. IEEE Test Conf., Paper 13.4, pp 397–401.

ENTROPY

2.19 Dussault J.A., 1978, "A testability measure", Proc. IEEE Semiconductor Test Conf., pp 113–116.

OTHER REFERENCES

2.20 Bennetts R.G., 1975, "On the analysis of fault trees", IEEE Trans. on Reliability, R–24, pp 175–185.

2.21 Dejka W.J., 1977, "Measure of testability in device and system design", Proc. 20th Midwest Symposium on Circuits and Systems, pp 39–52.

2.22 Keiner W., and West R., 1977, "Testability measures", Proc. Autotestcon, pp 49–55.

2.23 Bennetts R.G., 1978, "Algebraic models for clocked flip–flops", Electronic Eng., August, pp 59–63.

2.25 Danner F., and Consolla W., 1979, "An objective PCB testability rating system", Proc. IEEE Semiconductor Test Conf., pp 23–28.

2.26 Goel D.K., 1981, "A measure for test length and difficulty of generating tests for digital logic circuits", ITT LSI Systems Support Centre, Report No. 81–06, 20th Aug.

2.27 Takasaki S., et al., 1981, "A calculus of testability measure at functional level", Proc. IEEE Test Conf., pp 95–101.

2.28 Bennetts R.G., 1982, "Introduction to digital board testing", Crane–Russak Ltd. (New York), Edward Arnold (London).

2.29 Fong J.Y.O., 1982, "A generalised testability analysis algorithm for digital logic circuits", Proc. IEEE Symposium on Circuits and Systems, pp 1160–1163.

2.30 Goel D.K., and McDermott R.M., 1982, "An interactive testability analysis program – ITTAP", Proc. IEEE 19th Design Automation Conf., pp 581–586.

2.31 Argrawal V.D., and Mercer M.R., 1982, "Testability measures – what do they tell us?", Proc. IEEE Test Conf., Paper 13.3, pp 391–396.

2.32 Kirkland T., and Flores V., 1983, "Software checks testability and generates tests of VLSI design", Electronics International, Vol. 56, No. 5, March 10, pp 120–124.

3

Structured design techniques and self test

There have been many proposals for design techniques that produce testable logic circuit designs. Many of these proposals have related only to combinational circuits or have incurred too high a penalty in terms of requirements for additional primary inputs, primary outputs, and gates. Certain techniques have emerged as durable and viable however and it is these techniques that are described in this chapter. In particular, the chapter will cover the general principle of a structured approach to testable design called 'Scan-In, Scan-Out' or, more commonly, just 'Scan'.

The initial practical development of scan design techniques was carried out largely by electronics systems companies with captive IC manufacturing capabilities, and not by the IC makers themselves. This had two effects. The first was that individual companies produced their own scan design variants and the second was that implementation of scan techniques at PCB level, using standard catalogued devices, was considered to be too costly.

The situation is now changing. In the first place, more and more designers are able to influence, and even dictate, how their designs are to be implemented directly into silicon through custom-LSI and gate-array techniques. These designers need to know about scan design: its principles, advantages and penalties. In addition, IC manufacturers are beginning to produce standard devices with in-built scan facilities, the first of which was announced almost as this book went to press. This is the Am29818 Serial Shadow Register device produced by AMD Inc., discussed more fully in Section 5.1 of Chapter 5 and described in the Appendix.

For these reasons therefore, a discussion of scan design is appropriate in this book and this chapter describes three variations on the implementation of scan techniques: Scan Path, Random Access Scan and Level-Sensitive Scan Design (LSSD). Each of these techniques has been developed by large companies and LSSD in particular, developed by IBM, has received considerable attention.

A fourth less rigorous approach called Scan Set is also available and is described in Section 5.1 of Chapter 5. Unlike LSSD for example, Scan Set is less dependent on specific latch or flip-flop designs and more amenable as a general purpose

46

implementation. This is the reason for delaying the discussion until Chapter 5. Nevertheless, the reader should note the name now and, essentially, read the description of Scan Set in Chapter 5 as an extension of the more rigorous techniques about to be described.

The chapter concludes with a discussion of the role of self test and describes the basic Signature Analysis technique. This is followed by the more sophisticated Built-In Logic Block Observation (BILBO) approach which combines the scan concept with a self-test facility.

3.1   PRINCIPLE OF SCAN DESIGN: SCAN PATH

Scan design is a deliberate attempt to reduce the complexity of the test-generation problem for logic circuits containing stored-state devices and global feedback. The philosophy is one of 'divide and conquer' and is described with reference to the general model for a clocked (synchronous) logic circuit shown in Fig. 3.1.

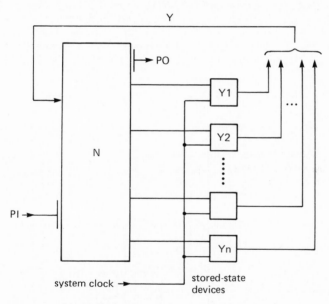

Fig. 3.1   General model for a synchronous circuit.

In this diagram, the major elements of the circuit have been identified as a combinational section N together with a bank of stored-state devices Y1, Y2, ...,Yn under the control of a system clock. Inputs to N consist of the primary inputs (PI) and the fed-back secondary-state variables Y. The primary outputs (PO) of the circuit are therefore a function of the present state of the primary inputs together with the current recorded output state of the stored-state devices. Likewise, the future state of the stored-state devices also depends on both the primary inputs and the current recorded state of the stored-state devices themselves.

It is this dependency of the future state on the present state that causes all the problems in test generation. The primary inputs are the only inputs over which the test programmer has direct control. Similarly, the primary outputs are the only outputs that can be observed directly. Control and observation of the stored-state devices is indirect through the combinational section of the circuit. The problem is – which section do we test first given that neither section is directly controllable or observable and that the sections are mutually dependent on each other for correct operation?

The scan-design technique provides a solution to this problem by reducing the complexity of the circuit structure. The principle of the technique is to provide additional facilities within the circuit, such that:

(a) the stored-state devices can be tested in isolation from the rest of the circuit;
(b) the future state of the secondary variables can be set to any desired set of values independent of their present values;
(c) the outputs of the combinational circuit that drive into the stored-state devices can be observed directly.

The method by which this is achieved is to establish a scan path through the stored-state devices, as shown in Fig. 3.2.

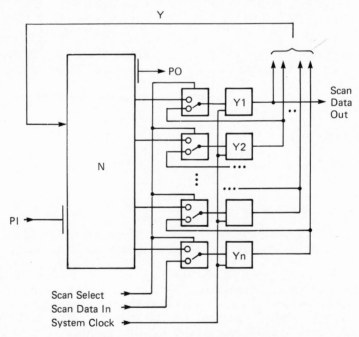

Fig. 3.2 Principle of scan path.

Effectively, each stored-state device is now preceded by a 2-way switch (multiplexer) under the control of a common Scan Select signal. When Scan Select is off, the multiplexers connect the outputs from the combinational logic through to the input sides of the stored–state devices, i.e. the circuit works in its normal mode. When Scan Select is on, the stored–state devices are reconfigured into an isolated serial–in, serial–out shift register. The serial data input is called Scan Data In and the serial data output is called Scan Data Out. In the scan mode therefore, the stored–state devices can be preset to any particular set of values simply by placing the values in sequence on the Scan Data In input and clocking the shift register with the System Clock.

The testing strategy is now as follows.

STEP 1   Select the scan–path mode, i.e. stored–state devices reconfigured into a shift register. Test the status and operation of each stored–state device using the Scan Data In, Scan Data Out and System Clock facilities. Suitable tests for a scan–path register are as follows:

(a) Flush test In this test, all stored–state devices are initialised to 0 and a single 1 is clocked through from the Scan Data In input to the Scan Data Out output using the scan–path clock. The procedure can be repeated with a single 0 flushed through a background of 1s. This sequence checks the ability of each stored–state device to assume both a 0–state and a 1–state.

(b) Shift test. In this test, the sequence 00110011... is shifted through the register. This sequence exercises each stored–state device through all combinations of present state and future state.

STEP 2   Determine a set of tests for the combinational logic, assuming

(a) total control of all inputs (primary and from the stored–state devices);
(b) direct observability of all outputs (primary and to the stored–state devices).

STEP 3   Apply each test in the following way:

STEP 3A   Select scan–path mode. Preload the stored–state devices with test input values and establish additional test input values on the primary inputs.
STEP 3B   Select normal mode. The steady–state output response of the combinational logic can now be clocked into the stored–state devices.
STEP 3C   Return to scan–path mode and clock out the contents of the stored–state devices. Compare these values, plus the values directly observable on the primary outputs, with the expected fault–free response.

The 'divide–and–conquer' philosophy of the scan design approach can now be seen more clearly. Rather than test the circuit as a single entity, the addition of the scan path allows each major segment to be tested separately and in a procedural manner. Furthermore, if we assume a standard test for the stored–state devices (Step 1 above), the only test–generation problem is to generate tests for the combinational segment. This problem has been well researched and a variety of programmable procedures exist. The next chapter describes one such procedure known as RAPS/PODEM.

One immediate problem with the circuit shown in Fig. 3.2 (and Fig. 3.1) is that the possibility of a race condition exists as output data from the combinational circuit is clocked into the stored–state devices. The problem is caused combinational circuit is clocked into the stored–state devices. The problem is caused by different response times in the stored–state devices. A device that responds quickly can modify an output from the combinational logic before a slowly responding device has had time to react. This is a classical problem with synchronous design and is usually solved by using master–slave devices. Fig. 3.3 shows a suitable design for a master–slave D–type flip–flop that incorporates the scan–path facility.

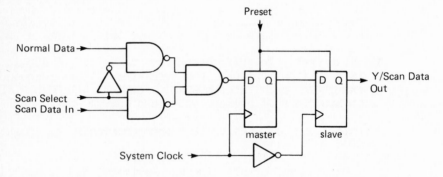

Fig. 3.3 Master–slave D–type flip–flop with scan path.

This design could be constructed from off-the-shelf standard devices but the advantges of the scan facility could be outweighed by the extra cost of components. The next sections describe more-specific implementations of scan that call for special-purpose designs of the stored–state devices.

3.2 RANDOM ACCESS SCAN

The principle objective of the Random Access Scan (RAS) design technique is to allow each stored–state device to be separately addressed in order that it can be independently set or preset, or its output value observed. Figure 3.4 illustrates the principle of the technique and Fig. 3.5 shows one particular implementation of the stored–state device: in this case, a preset/clear addressable latch.

In Fig. 3.4, each latch is individually selected via the

decoded output of the scan-address register. (Note, this register
is optional. It serves to reduce the number of primary inputs
required but increases cost and test-application time.)

The various modes of operation of the preset/clear addressable
latch are as follows.

(a) To clear the latches. Both CLR and PR are set low (logic 0)
and the system clock (CLK) set high (logic 1). This sets the
Q output to logic 0. Following the clear operation, CLR is
taken high and PR left low.

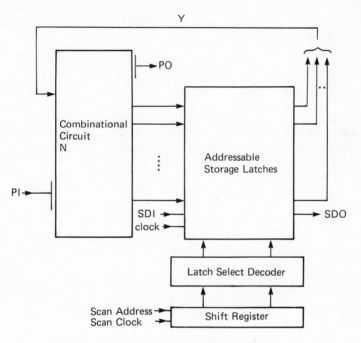

Fig. 3.4 Random-access scan.

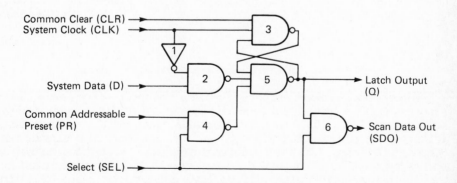

Fig. 3.5 Preset/clear addressable latch.

52

(b) Normal latch operation, responding to changes on the system data (D) line. Assuming CLR = 1 and PR = 0, any change in the value of D is passed straight through G5 of the cross-coupled NAND gate pair when CLK = 0. The actual value latched is the last value on D as CLK returns high.

(c) To preset a latch. Following a clear operation, individual latches can be preset by selecting the latch and setting PR high, with CLR and CLK held high. The Q output of the selected latch will go high and remain so even when the PR line is taken low again.

(d) The value on the Scan Data Out (SDO) line is observable when the latch is selected.

Figure 3.6 shows an alternative implementation of an addressable latch for use in a RAS environment. Normal operation requires the Scan Clock (SCLK) to be held low, in which case changes in System Data (D) are transferred through to Q when the System Clock (CLK) is low. The last value on D is latched as CLK goes low to high. Scan operation is controlled similarly by the Scan Clock (SCLK) and requires the System Clock (CLK) to be held high. When the latch is selected, the latch output can be set to the value on Scan Data In (SDI) or the latched value observed on Scan Data Out (SDO).

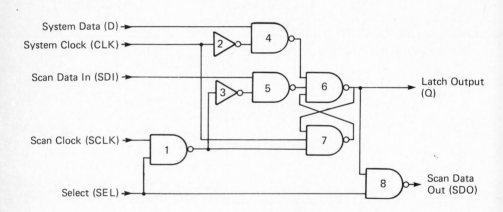

Fig. 3.6 Polarity-hold addressable latch.

Random Access Scan differs in one respect from the basic scan-path philosophy insofar as it does not contain a scan path as such. Individual SDO lines are normally high (for non-addressed latches) and can be tied together and brought out as a single Scan Output (SO) line. If the selected latch has Q = 0, then there is no change in the observed SO value. If the selected latch has Q = 1, then the SDO value will go low, pulling the main SO line low. The SDO values of all latches are determined by cycling through all addresses.

Obviously, the major penalty with the Random Access Scan approach is the amount of time necessary to set the test input values into the latches and subsequently to observe the latched responses. Also the overhead in additional gates is relatively high. The next section describes a more direct implementation of the scan design approach which does not carry such onerous penalties in test-application times or additional gates.

3.3 LEVEL-SENSITIVE SCAN DESIGN

There are two fundamental constraints to the Level-Sensitive Scan Design (LSSD) technique. The first is that all changes to the state of the circuit are controlled by the level of a clock control signal, rather than by the edge of the clock. Furthermore, the steady-state response to a change of value on a primary input is independent of the propagation delay of gates and interconnect elements within the circuit. The response is also independent of the order of input-value changes in the event of simultaneous multiple changes. This is the property of 'level-sensitivity', designed to reduce the dependency of the circuit on its ac parameters, such as degraded rise and fall times, degraded propagation delays, or other faults that have the potential of introducing race or hazard conditions. Overall therefore, the potential effect of failure mechanisms that cause timing faults is reduced.

The second constraint of LSSD circuits is that the circuit should possess the scan-path property. In this case, this is achieved by using a special purpose stored-state device called a polarity-hold Shift Register Latch (SRL), shown in block diagram form in Fig. 3.7 and NAND-gate equivalent in Fig. 3.8.

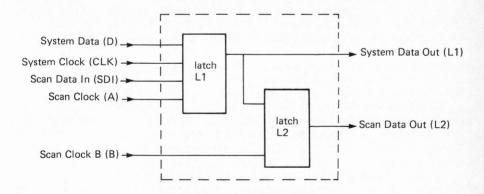

Fig. 3.7 Polarity-hold shift register latch.

3.3.1 SRL operation and LSSD configurations

The SRL contains two cross-coupled NAND gate latches, L1 and L2. L1 constitutes the normal stored-state holding device with System Data (D) and System Clock (CLK) inputs and L1 System Data output. Normal latch operation requires Scan Clock A to be held low. Scan

Clock B is also held low during normal system operation. Latching of System Data then occurs as the System Clock is returned low from an active high value.

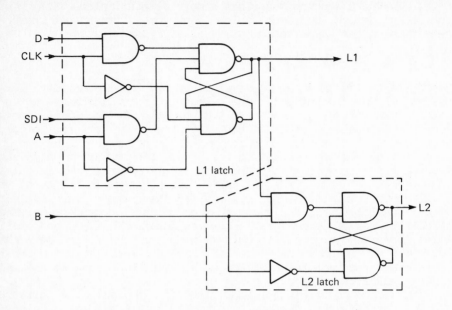

Fig. 3.8 NAND–gate implementation of an SRL.

To operate the SRL as part of a scan path, Scan Clock A is set to 1. This enables data on the Scan Data In (SDI) input to be latched directly into latch L1. Scan Clock A is then returned low (to latch the value into L1) and Scan Clock B raised high. This causes transfer of the latched L1 value into latch L2 with permanent latching in L2 as B is returned low.

In a practical LSSD circuit, the SRLs are connected permanently to form a scan–path shift register by connecting the L2 output (Scan Data Out) of one SRL to the Scan Data In input of another SRL. The two scan clocks, A and B, are common to all SRLs. Figure 3.9 shows one particular way of implementing an LSSD circuit based on SRLs. This configuration is called the 'double–latch'.

In the double–latch configuration, the L1 output of each SRL is not used – only the L2 output. This output serves both as the System Data output (the stored–state values, Y) and as the Scan Data Out output connected through to the following SRL's Scan Data In input. In effect, each SRL operates in a master–slave mode with data transfer occuring under the control of either System Clock and Scan Clock B (for normal system operation) or Scan Clock A followed by Scan Clock B (for scan–path operation). Race–less behaviour is therefore guaranteed in either mode of operation. The name 'double–latch' comes from the fact that, in any SRL, both latches are in the system path.

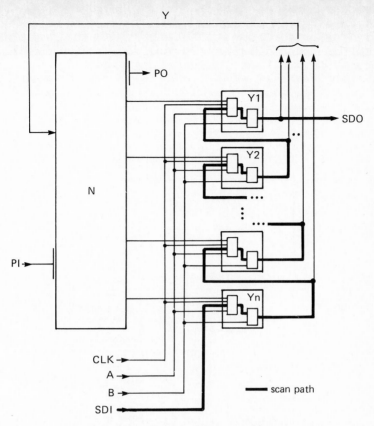

Fig. 3.9 Double-latch LSSD configuration.

Figure 3.10 shows an alternative way of using SRLs in an LSSD environment, called the 'single-latch' configuration. This configuration makes use of the L1 output as the system output and avoids the potential race condition by partitioning the combinational logic into two disjoint sets, denoted N(1) and N(2) in Fig. 3.10. System clocks into the N(1) and N(2) SRLs are denoted CLK(1) and CLK(2) respectively. The outputs of the SRLs associated with N(1) become the secondary variable inputs to N(2), and vice versa. System operation is controlled by the two system clocks, CLK(1) and CLK(2), which operate in such a way as to ensure that only one clock is active (high) at any one time, i.e. CLK(1) and CLK(2) are non-overlapping. In this way, potential race conditions are avoided. The name 'single-latch' comes from the fact that only one latch (L1) is used in the system path.

The essential difference between the double-latch and single-latch configuration lies in the speed with which the circuit primary outputs can change as a result of primary input and clock changes. The double-latch system requires two independent and non-overlapping clocks (CLK and B) to change before signal-value changes can be propagated through the L1 and L2 latches and thence through the combinational circuit N to produce a stable primary output value. The single-latch configuration on the other hand only requires the appropriate single clock to change (CLK(1) or CLK(2)) to cause propagation through the L1 latch before the appropriate combinational circuit outputs (N(2) or N(1) respectively) can stabilise. In both cases, the fastest operating speed is governed by the propagation delay of the combinational logic circuit. If this delay is denoted by N-delay(max), then the maximum clock rate on the system clock, CLK for double-latch and CLK(1) or CLK(2) for single-latch, is given by:

$$CLK(max) < N\text{-}delay(max)$$

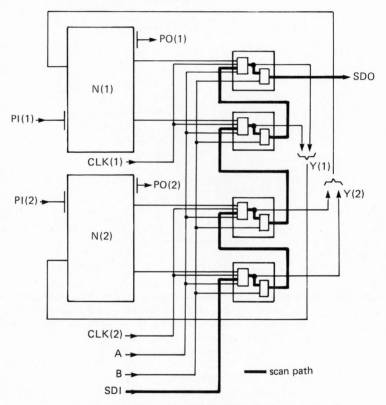

Fig. 3.10   Single-latch LSSD configuration.

3.3.2   Use of the L1/L2* SRL

A disadvantage of the single-latch configurations based on the SRL of Fig. 3.8 is that the L2 latch has no role to play in system operation. In that sense, the L2 latches are redundant and represent a high testability overhead in terms of extra silicon. A recent modification to the basic SRL latch is designed to reduce this penalty by making the SRL more versatile. The modified latch is called the L1/L2* SRL and is shown in gate-equivalent form in Fig. 3.11.

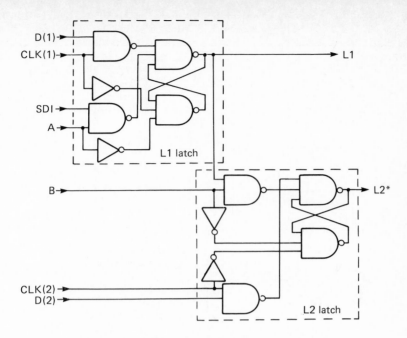

Fig. 3.11 L1/L2* shift-register latch.

The main difference between the basic L1/L2 SRL and the
modified L1/L2* SRL occurs between the L2 latches. The L2* latch
contains facilities to allow an alternative system data input,
D(2), to be latched in under the control of a separate system
clock, CLK(2). The original system data input – D in Fig. 3.8,
D(2) in Fig. 3.11 – is still available and under the control of
the original system clock, now called CLK(1). CLK(1) and CLK(2)
are non-overlapping clock signals. Scan-path facilities are the
same in both cases and scan data transfer is controlled as before
by the scan clocks A and B.

The L1/L2* latch can be used in a single-latch configuration
in such a way as to allow the system output to be taken from
either the L1 output or the L2 output. This is demonstrated in
Fig. 3.12.

In this circuit, the data outputs from N(1) are taken to the
D(1) inputs of the L1 latches. The data outputs from N(2) are
taken to the D(2) inputs of the L2* latches. Both latches, L1 and
L2*, are therefore used as part of the system, unlike the
single-latch configuration based on the L1/L2 SRL. Note that
system outputs to the same combinational circuit, N, can be taken
from either L1 or L2 but not from both simultaneously.
(Figure 3.12 shows L1 outputs used as inputs to N(2) and L2
outputs used as inputs to N(1).) There are two reasons for this
restriction. The first is that to take both L1 and L2* outputs
back to the same combinational circuit effectively removes the
master-slave relationship between L1 and L2*, in any single SRL,
caused by the alternate paths through N(1) and N(2).

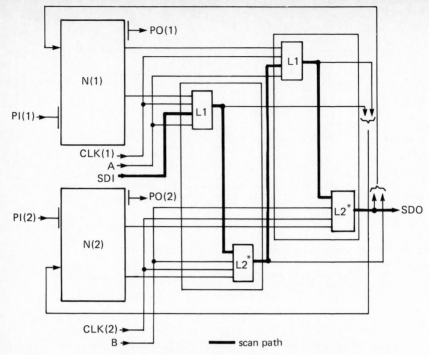

Fig. 3.12   Single-latch LSSD based on the L1/L2* SRL.

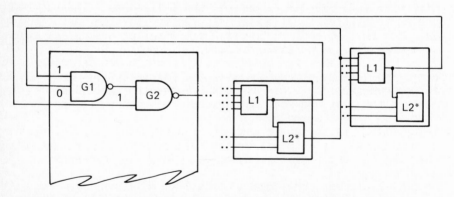

The 101 values shown on gates G1 and G2 are a necessary test for the two gates.
These values cannot be set up on the L1, L2* outputs of the two latches.

Fig. 3.13   Example of untestable single-latch configuration.

The second is that, in the test set-up mode, it may not be possible to establish the required test data on L1 and L2* outputs because the transfer of values from L1 to L2* outputs and from L2* to the next L1 is controlled by two separate non-overlapping clocks. This means that either the value on L2* is always matching that on L1 or that the value on the next L1 is always matching that on the previous L2*. This places a restriction on the exact sequence of 0s and 1s and can cause conflict as illustrated in Fig. 3.13.

3.3.3 Design rules

The previous sections have outlined various ways in which the scan-path philosophy can be implemented using the SRL device. There are obviously other variations to suit outside constraints and to accommodate the ingenuity of circuit designers. In practice however, design disciplines such as LSSD must be supported by a design automation system which includes a rigorous check for any violations to the basic design rules. Design rule checking, as this is called, is an essential feature of LSSD and this section describes the basic rules for producing an LSSD system. Some of the rules are self-evident: others are designed to preserve either the level-sensitive property or the scan property. The rules are as follows.

Rule 1   All stored-state devices are SRLs.

Rule 2   SRLs are controlled by two non-overlapping clocks such that:

(a) the L1 or L2/L2* output of SRL(1) can be used to gate a clock C to produce a gated clock, C(G). C(G) can then be used to clock another latch, SRL(2), provided SRL(1) is not being clocked by C;
(b) subject to this restriction, the outputs of SRL(1) may feed the data inputs of SRL(2).

Rule 3   It must be possible to identify a set of SRL clocks that are directly controllable. This means that:

(a) all clock inputs can be held inactive independently;
(b) any single clock can be made active whilst the others are maintained in their inactive state.

Rule 4   Clock primary inputs can only be connected to SRL clock inputs. They cannot be connected to SRL data inputs, either directly or through the combinational logic circuit.

Rules 1-4 constitute a check for the property of level sensitivity The following two rules constitute a check for the property of scan design.

Rule 5   All SRLs are permanently connected to form a scan-path register with a scan-in primary input, scan-out primary output and accessible scan control clocks.

Rule 6   There must exist a circuit configuration state, directly controllable from the primary inputs and called the 'scan' state, such that:

(a) all SRLs are connected and available as a scan path;
(b) all SRL clocks, except the shift clocks A and B, can be held inactive;
(c) any and each shift clock, A and B, can be made active independently.

Circuits which conform to Rules 1-6 inclusive are said to be LSSD circuits.   The single-latch and double-latch configurations described earlier are simple examples of LSSD circuits. Undoubtedly, more complex configurations can be designed but, as with most testability constraints, the objective is to keep the design structure simple.

3.4 PENALTIES AND ADVANTAGES OF SCAN DESIGN

The major penalties of imposing a scan design policy on the design of a logic system can be categorised as follows.

(a) Additional physical requirements. The most immediate penalty is the requirement for extra primary input/output access points and logic devices.
Specifically for LSSD, the number of additional input/output points is four (SDI, A, B, SDO) and the percentage increase in the number of gates has been estimated to lie between 4% and 20%, according to how the gates are judged to contribute to the normal operation of the circuit.
The net result of this additional circuit facility is higher parts-count and higher connect-count leading to higher manufacturing cost and lower reliability.

(b) Increased test application time. The testing strategy associated with a scan design policy requires the circuit to cycle from scan-path mode to normal mode and back again as each test is loaded and applied. The serial nature of the strobe-in strobe-out mechanism can create long test-application times - long in terms of the amount of time necessary to apply all the tests to the combinational circuit as required by a fault location procedure based on a hand-held walk-back guided probe. (Operator fatigue starts to become a problem when the signature collection time reaches the order of 30 secs. Given that the signature may be collected three times during this period to ensure consistency and therefore accuracy, this puts a realistic limit of 10 secs duration to apply all the tests.)
It has been suggested that the test-application time could be reduced by designing a parallel-load, parallel-read scan path but any advantage gained would be more than offset by the additional burden of input/output requirements.

(c) Inability to test the circuit at full operational speeds. Certain fault conditions, such as dynamic timing interaction faults caused by a degradation of some propagation-delay parameter, are only detected by exercising the circuit at or near its maximum operating speed. The 'stop-go' nature of testing a scan design prevents any real attempt to test for such faults.

(d) Constraints on designer freedom. A scan design policy places constraints on the freedom of choice traditionally associated with logic design. In particular, the designer is not able to design asynchronously to obtain a speed increase. Neither can the designer select from a wide range of off-the-shelf stored-state devices: RAS and LSSD are quite dictatorial about the nature of the stored-state devices. Scan design can therefore seem to reduce the processes of logic design to fairly simple mechanical processes with little scope for real creativity. This is not the case however. The inventiveness, and therefore the fun of logic design, is still present partly because not all circuits are suitable candidates for a scan-path design, e.g. microcomputer boards, and partly because there is still a considerable amount of research to carry out in the area of synthesis techniques to turn a design specification into a scan implementation. Currently, very little work has been done in this area.

To offset these penalties, the following advantages are associated with scan designs.

(a) Design validation and timing analysis. The ability to carry out automatic checks for violations of a set of design rules means that the processes of design validation (zero-delay fault-free logic simulation) and timing analysis are simplified. LSSD is specifically aimed at reducing the dependency of system operation on ac parameters, such as clock edge rise or fall times, which are difficult to simulate in a design environment and difficult to monitor in a manufacturing environment.

(b) Test-pattern generation. For scan-designed circuits, algorithmic test-pattern generation is required only for the combinational circuits. This requirement is well understood and many practical procedures exist to produce an acceptable solution. The next chapter describes one such procedure called RAPS/PODEM.

(c) Test-pattern evaluation. The reduction of the complexity of the test-generation process carries with it an added bonus for test-pattern evaluation, if the evaluation is carried out by fault simulation. The bonus is that the simulator need only evaluate the fault-cover of tests applied to combinational circuits. This means that the simulator can be specially designed and therefore will be faster in operation than the more general-purpose fault simulators. This speed increase also allows the simulator to be used as an integral part of the test-generation process to evaluate the full fault-cover of each test as it is generated. The RAPS/PODEM procedure, described in the next chapter, makes use of this feature.

(d) Test application and fault diagnosis. Partitioning of the circuit into a scan path and a combinational section eases the problem of ensuring correct fault diagnosis both in the manufacturing environment and the field-servicing environment. The ability to open the feedback paths (by operating the circuit in the scan-path mode) removes the major cause of incorrect or ambiguous diagnosis. Walk-back guided probes can be used to locate precise sources of failure both in the stored-state devices (circuit in scan-path mode) and in the combinational section (circuit in normal mode).

An advantage in the area of test application is that the circuit presents a uniform and well-defined interface to the tester. Construction of the complete test program – set-up instructions, scan-path tests, combinational circuit tests – is not difficult. Indeed, it could be as simple as linking together parameterised program modules to carry out each of these tasks. A spin-off advantage from the inherent modular structure of the test program is that separate modules can be easily replaced with updated versions. This in turn alleviates the perennial problem of upgrading test programs as modifications are made to the design. In the field, such modifications, called Engineering Change Orders (ECOs) or something similar, require careful handling in terms of upgrading the test program and associating the correct program with a particular revision level of the design. A modular test program assists in reducing the problems of managing ECOs.

In summary, scan design techniques offer significant advantages in all areas of testing and yet still present the designer with some interesting challenges. The major penalty appears to be the need for extra primary input/output access and gates but this penalty is common to all design-for-test techniques. As always, the costs of testing will determine which way the balance will tip.

3.5 SELF-TEST TECHNIQUES: BILBO

The use of built-in-test facilities is a time-honoured method for instilling confidence about the correctness of operation of a logic circuit. Probably the best known approach is through the concurrent (on-line) use of a variety of error-detecting, error-correcting and even self-checking codes, e.g. simple parity, Hamming, m-out-of-n, Berger, etc. The attributes of these codes are well known and have been described in many textbooks. The purpose of this section is to discuss an approach to self-test called Signature Analysis (SA), based on the use of a source of test stimuli actually built onto a chip itself or placed on a printed-circuit board. The idea is not new but has recently been revived, initially as a way of improving the chances of a field-service engineer locating a fault on a board, and hence carrying out an effective repair, without returning the faulty board to a central repair depot.

The reason for discussing signature analysis is that there has recently been an interesting development which seeks to combine the elements of signature analysis with those of scan design. This approach is termed Built–In Logic Block Observation (BILBO) and is also described in this section.

3.5.1 Use of CRC signatures

A standard method for locating the source of failure in a digital circuit is to use a hand–held guided probe to look at the status (good or bad) of the binary information being generated on a particular circuit node as the circuit is excited by a pre–defined set of test stimuli. It is common to apply some form of data compression to the expected fault–free sequence of 0s and 1s on each circuit node, the most common compression technique being the Cyclic Redundancy Check (CRC) form. The CRC form is based on the use of a Linear Feedback Shift Register (LFSR), usually 16–bits long, as shown in Fig. 3.14.

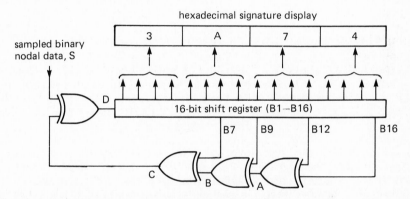

The feedback points in this register follow those advocated by Hewlett-Packard.

Fig. 3.14 Linear feedback shift register.

Binary data are sampled from the logic circuit at defined strobe (test) times and clocked into the shift register, the initial contents of which are zero. The contents of the register are influenced not only by the next sampled value but, by virtue of the feedback structure, by the current contents of the register. In this way, any corruption of the sampled bit stream, caused by a generated or transmitted fault condition on the sampled node, causes a corresponding corruption in the contents of the shift register. At the end of the sampling period, the accumulated contents of the register, i.e. the nodal signature, are displayed (usually in hexadecimal format) and compared with the expected fault–free signature. A match indicates that the node is fault–free; a mismatch indicates that the data on the node are corrupt in some way. For CRC signatures, the probability of a corrupt bit stream generating the same signature as the fault–free bit stream is extemely low, quickly approaching 1 in 2^n as the length of the bit stream exceeds the length n of the shift register. In practice, this limitation is not significant and many testers now make use of guided–probe fault–locating systems based on a knowledge of a pre–stored set of expected fault–free CRC signatures for all signal–carrying circuit nodes. Figures 3.15–3.16 illustrate the technique applied to the master–slave D–type with scan path discussed earlier (Fig. 3.3).

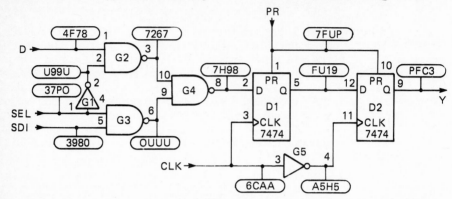

Fig. 3.15   Master–slave D–type with fault–free signatures.

Figure 3.15 shows the circuit, this time with 4–character hexadecimal CRC signatures associated with each output node. These signatures are derived from the sequence of test patterns defined in Fig. 3.16 which provide 100% fault–cover for all single s–a–1, s–a–0 faults. The hexadecimal character set is non–standard – 0,1,2,3,4,5,6,7,8,9,A,C,F,H,P,U – to allow easier readability on 7–segment display systems. (This character set is advocated by the Hewlett–Packard Corporation – hence the use of the H and P characters!) Conversion of the bit patterns into 4–character hexadecimal signatures is demonstrated with the G4.8 sequence defined in Fig. 3.16 and based on the feedback characteristics shown in Fig. 3.14. The detail is shown in Fig. 3.17.

Consider now what happens if, for example, an open circuit occurs on the connection between G3 and G4. Assuming TTL logic, the binary sequence on G3.6 will be correct but the binary sequence on G4.9 will be all–1s. As the tests are applied, then so the faulty all–1 sequence will cause incorrect sequences on succeeding nodes along the sensitive path through to the primary output; in this case, nodes G4.8, D1.5 and D2.9. Figure 3.18 shows the CRC signatures corresponding to the modified behaviour of the circuit in the presence of the fault.

Detection of the fault condition is by tester observation of the incorrect binary sequence on the primary output, Y. Location of the fault can be carried out by prompting an operator to work back with a data–collecting probe from the observed faulty primary output until the source of the faulty sequences is found. The collected data are strobed into the CRC shift register and, on completion of the set of tests, the contents of the register are displayed. The operator can now compare the displayed value with the expected value and decide whether the source of the fault has been identified, or whether further probes should be made.

	PI set-up				Internal values			Clock	Clocked values		Comment
	PR	SEL	D	SDI	G2.3	G3.6	G4.8	CLK	D1.5	D2.9 (Y)	
*	0	0	0	0	1	1	0	0	1	1	Initialise
	1	0	0	0	1	1	0				Release PR
*								1	0	1	
*								0	0	0	
	1	0	1	0	0	1	1				
*								1	1	0	
*								0	1	1	
	1	1	1	0	1	1	0				Tests for:
*								1	0	1	G1-G4 s-a-1, s-a-0
*								0	0	0	PR s-a-0
	1	1	0	1	1	0	1				CLK s-a-1, s-a-0
*								1	1	0	
*								0	1	1	
	1	1	0	0	1	1	0				
*								1	0	1	
*								0	0	0	
	1	0	0	1	1	1	0				
*								1	0	0	
*								0	0	0	
*	0	0	0	1	1	1	0		1	1	
*								1	1	1	PR s-a-1
*								0	1	1	

Assumptions

(i)   PI set-up values remain constant during clock sequence. i.e. a blank entry denotes no change from the previous value.

(ii) The clock sequence is applied some time after the internal values have had time to stabilise following a change on the PIs.

(iii) Strobe points (for data collection) occur every time the CLK changes value with enough delay to ensure stable values on the output of D1 (as clock goes high) and D2 (as clock returns low). Strobe points are marked * in the table above.

Example

The binary sequence expected on G4.8 is given by:

0 0 0 1 1 0 0 1 1 0 0 0 0 0 0 0

First strobe point   Last strobe point

Fig. 3.16 Test patterns for Fig. 3.15.

The part of the circuit containing the failure can be identified when one of the following conditions is satisfied.

Condition 1   A device is found with a faulty signature on an output pin but with correct signatures on all the input pins which can logically affect the binary value on the output pin.

Condition 2   A connection is found with a correct signature at one end (the signal source end) and an incorrect signature at the other end, or one of the other ends in the case of a fanout connection.

For Condition 1, the diagnosis points to the device itself as the probable cause of failure but there is also a chance that the connection leading from the device output is faulty. For Condition 2, the diagnosis points to some problem with the connection, usually an open circuit.

Starting Values S A B C D	Clock	Shift-register contents														
		B1	B2	B3	B4	B5	B6	B7	B8	B9	B10	B11	B12	B13	B14	B15	B16
		0	0	0	0	0	0	0	0	0	0	0	0	0	0	0	0
0 0 0 0 0																	
		0	0	0	0	0	0	0	0	0	0	0	0	0	0	0	0
0 0 0 0 0																	
		0	0	0	0	0	0	0	0	0	0	0	0	0	0	0	0
0 0 0 0 0																	
		0	0	0	0	0	0	0	0	0	0	0	0	0	0	0	0
1 0 0 0 1																	
		1	0	0	0	0	0	0	0	0	0	0	0	0	0	0	0
1 0 0 0 1																	
		1	1	0	0	0	0	0	0	0	0	0	0	0	0	0	0
0 0 0 0 0																	
		0	1	1	0	0	0	0	0	0	0	0	0	0	0	0	0
0 0 0 0 0																	
		0	0	1	1	0	0	0	0	0	0	0	0	0	0	0	0
1 0 0 0 1																	
		1	0	0	1	1	0	0	0	0	0	0	0	0	0	0	0
1 0 0 0 1																	
		1	1	0	0	1	1	0	0	0	0	0	0	0	0	0	0
0 0 0 0 0																	
		0	1	1	0	0	1	1	0	0	0	0	0	0	0	0	0
0 0 0 1 1																	
		1	0	1	1	0	0	1	1	0	0	0	0	0	0	0	0
0 0 0 1 1																	
		1	1	0	1	1	0	0	1	1	0	0	0	0	0	0	0
0 0 1 1 1																	
		1	1	1	0	1	1	0	0	1	1	0	0	0	0	0	0
0 0 1 1 1																	
		1	1	1	1	0	1	1	0	0	1	1	0	0	0	0	0
0 0 0 1 1																	
		1	1	1	1	1	0	1	1	0	0	1	1	0	0	0	0
0 1 1 0 0																	
		0	1	1	1	1	1	0	1	1	0	0	1	1	0	0	0
		(7)				(H)				(9)				(8)			

Fig. 3.17 CRC signature for node G4.8

Fault locating techniques based on the use of CRC signatures and guided probes work well for acyclic circuits, i.e. circuits which do not contain global feedback paths which create a closed loop between physical devices.

For circuits which do possess global feedback paths, consider the effect of a fault originating within the loop. It is possible for the fault effect to travel completely around the loop thereby corrupting all the CRC signatures within the loop. The problem then is to separate the cause from the effect – that is, to 'break' the loop – and this cannot be done without additional information about each node; namely, the earliest time (tester strobe point) at which the logical value on the node deviated from its correct value. One way to obtain this information is to return to the full binary sequence but even this approach will fail if all the nodes deviate at the same strobe point. In any case, reversion to using the full sequence is contrary to the original objective of compressing the data into a more manageable form.

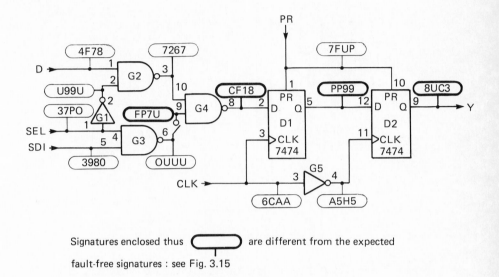

Fig. 3.18 Master–slave D–type with fault condition signatures.

A general solution to the so–called 'loop–breaking' problem has so far proved difficult to produce. Some proprietary solutions exist but these are not guaranteed to produce correct diagnosis every time. Practically, test programmers use various strategies to provide either more control or more observation of events within closed loops, as described later in Chapter 5. But, as we have already observed, the optimum solution from a testing viewpoint is to provide on–board or on–chip facilities to open–circuit feedback paths. This is one of the key features of self–testing based on the use of CRC signatures.

3.5.2 Signature Analysis

The term 'Signature Analysis' is used to describe a guided probe technique, based on the use of CRC signatures, for isolating faults that have developed in operational boards. It has been developed primarily to allow the service engineer to carry out field repairs. The only difference between Signature Analysis and what has been described in the previous section, is that the test stimulus is supplied by devices on the board, rather than from a tester, and the support equipment required is a portable instrument for capturing binary data at defined strobe times and displaying the CRC signature. Figure 3.19 illustrates the main features of a circuit designed to Signature Analysis standards.

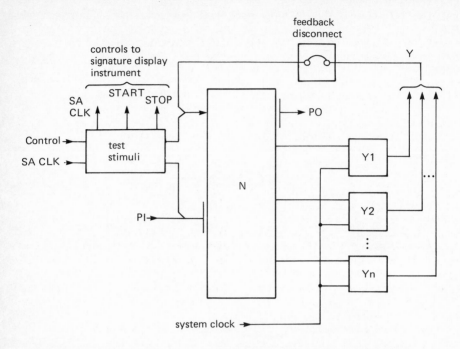

Fig. 3.19 Circuit designed for signature analysis.

There are two additional design features in Fig. 3.19. The first is the on-board source of test stimuli as a built-in test unit. This can either be some form of counter providing what is virtually pseudo-random test stimuli, or it can be a Read-Only Memory containing a more deterministic source of test stimuli.

The second design feature is a facility to break (open) all the global feedback paths (by switches, jumpers, or tristate buffers). This facility removes the loop-breaking problem.

The fault isolation strategy is now quite simple. The service engineer places a suspect board into its test mode of operation. In this mode, the test stimuli are applied to the circuit and the feedback paths are broken. As the stimuli are applied, each node will respond accordingly with a sequence of 0s and 1s. Working from an annotated circuit diagram or some form of troubleshooting flowchart, the engineer probes back through the circuit until an unambiguous diagnosis can be made (Condition 1 or Condition 2 as described in the previous section). In this way, the engineer is able to locate the fault, repair the board on site, and return it to operational status.

Signature Analysis has particular application to bus-structured designs, such as memory boards or microprocessor boards, because of the inherent ease of applying and monitoring the . tests. The tests can either be pre-loaded into the existing on-board RAM or they can be resident in part of an on-board ROM. Alternatively, a system ROM can be replaced by a test ROM. If the tests are resident then, effectively, the design can become self-testing with 'green-light, red-light' LED features built in. The application of this idea to board-level design is developed further in a later section of this chapter.

Returning to the principles of Signature Analysis, consider the requirements for the test stimuli. In practice, the test stimuli do not need to establish sensitive paths to the primary outputs of the circuit. A simpler requirement is to determine a set of input changes which succeed in changing the value of every circuit node at least once as the changes are applied to the circuit primary inputs. This requirement is called the 'node excitation' requirement. Input stimuli which have this property allow s-a-1 and s-a-0 faults to be located by probing every node and manually resolving cause-effect relationships. A possible conclusion to this observation is that the test stimuli source could be pseudo-random (p-r) rather than deterministic. A deterministic set of tests is derived in an analytical way and seeks to excite each circuit node in some pre-defined way. A p-r set of tests is simply a repeatable sequence of random input changes that might achieve some excitation objective but for which there is no guarantee. The tests in Fig. 3.16 are deterministic. They were derived with the collective objective of exciting all single nodal stuck-at faults and propagating these faults through to an observable primary output. Alternatively, the outputs of a 4-bit up-counter could have been used to stimulate the four control and data inputs PR, SEL, D and SDI.

The generation of tests by p-r techniques has received much attention by theorists and practitioners alike. Its attraction is the ease with which a large number of input stimuli can be defined. Its problem is the uncertainty over the precise fault cover of the stimuli. This uncertainty can be removed by fault simulation but fault simulation becomes prohibitively expensive very quickly as the number of input changes rises.

One area where p-r techniques do work well is in their application to purely combinational circuits. For large combinational circuits where exhaustive testing is not possible, a useful stratgy for generating tests is to investigate the fault cover of a small number of p-r patterns before using a more deterministic method, such as the D-algorithm. This strategy is the basis for the PODEM/RAPS approach described in the next chapter and is described in more detail there. The point about making these preliminary comments here is that p-r sequences have useful properties when applied to combinational circuits in terms of their ability to provide real test stimuli. A corollary to this property is that p-r sequences must also be a useful source of node excitation stimuli. This means that p-r generators could perform a useful role as the source of on-board test stimuli for circuits designed to Signature Analysis standards. The next question is - how can we generate p-r sequences? There are many solutions to this problem but one particular solution is to use a Linear Feedback Shift Register (LFSR) of the sort discussed earlier in connection with CRC signatures (Fig. 3.14 q.v.).

Figure 3.20 illustrates a simpler form of this type of shift register and shows the p–r sequence generated as the register is clocked.

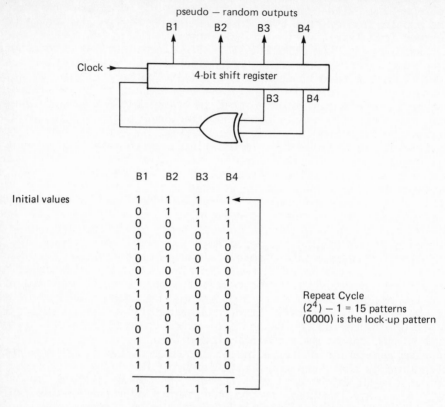

Fig. 3.20 Pseudo–random generator based on LFSR.

In terms of Signature Analysis, a p–r generator such as the one shown in Fig. 3.20 can be used as the test stimuli source of Fig. 3.19. Assuming some visibility of the outputs of the stored–state devices (Y in Fig. 3.19), the provision of the feedback disconnect facilities reduces the circuit behaviour to combinational plus clocked stored–state devices. Under these conditions, a p–r generator can be an effective source of test stimuli.

The next step is to consider the possibility of including data–collection facilities directly onto the board in order to build in a full 'test–and–compare' feature, i.e. self–test. This could be achieved by adding LFSRs configured as CRC signature generators to the circuit but the problem is where to place them and how many are needed. One solution could be to place separate LFSRs onto each secondary variable feedback path (Fig. 3.21); another could be to add a scan path to the circuit and use just a single LFSR (Fig. 3.22); yet another could be to use a parallel–input LFSR for collecting data from various key points such as the secondary variables (Fig. 3.23).

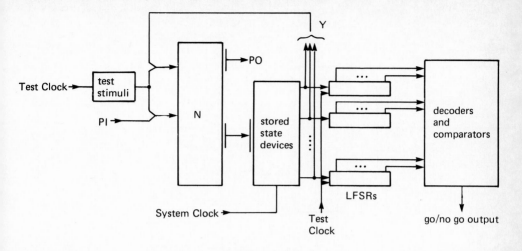

Test cycle: apply next test stimuli to N
             latch response into Y
             update LFSRs
             sample go/no go output on completion

repeat

Fig. 3.21 Self-test, scheme 1.

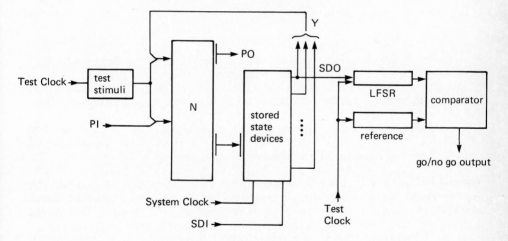

Test cycle: select normal mode
             apply next test stimuli to N
             latch response into Y
             select scan-path mode
             strobe out into LFSR
             sample go/no go output on completion

repeat

Fig. 3.22 Self-test, scheme 2.

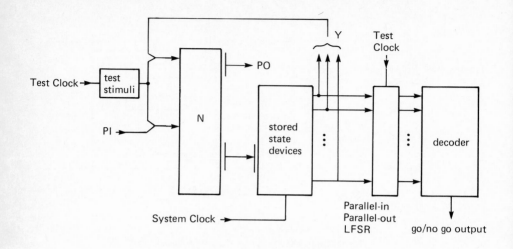

Fig. 3.23 Self-test, scheme 3.

These ideas, plus other variations, have stimulated the development of a universal self-test element – one that can perform all the functions of stimuli generation, CRC generation and even stored-state devices with scan path. An element that fulfils this requirement is the Built-In Logic Block Observation (BILBO) element. This element is now described.

3.5.3 Built-In Logic Block Observation element

The BILBO is a universal element for use in either a scan-path environment or a self-test (signature analysis) environment. The element is capable of performing a number of different functions according to the values placed on two mode-control lines. The general configuration of a 4-bit BILBO is shown in Fig. 3.24.

Essentially, the BILBO is a multiple-mode shift register with extra facilities to allow exclusive-OR feedback to occur. The four modes of operation are controlled by the the two mode-control lines, C1 and C2, and are shown in Figs. 3.25–3.28.

Figure 3.25 shows the reset mode, caused by setting C1 = 0, C2 = 1. Under these conditions, the input to each D–type latch becomes a 0 and, consequently, each latch can be initialised to 0.

Figure 3.26 shows the normal system operation mode corresponding to C1 = 1, C2 = 1. Each latch is independent and can be separately loaded (on the Z inputs) and read (on the Q outputs).

73

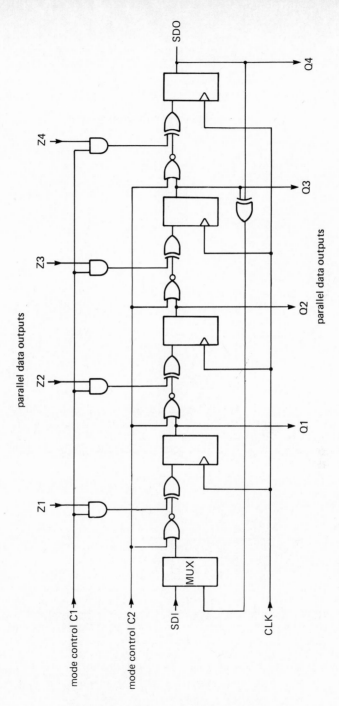

Fig. 3.24   Basic BILBO element

74

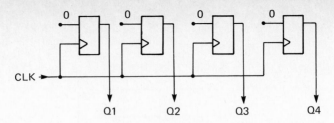

Fig. 3.25 BILBO reset mode: C1=0, C2=1.

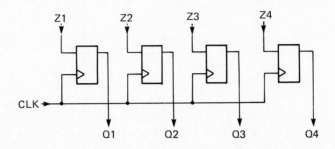

Fig. 3.26 BILBO normal latch mode: C1=1, C2=1.

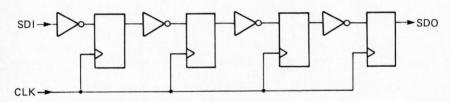

Fig. 3.27 BILBO scan path mode: C1=0, C2=0.

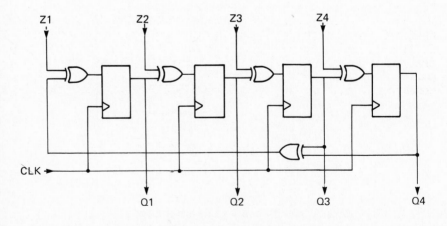

Fig. 3.28 BILBO LFSR mode: C1=1, C2=0.

The scan-path mode is shown in Fig. 3.27 and corresponds to C1 = 0, C2 = 0. The only unusual feature is the interstage inversion but this does not cause any real problems.

Finally, Fig. 3.28 shows the BILBO configured to act either as a p-r generator or as a CRC signature generator. This corresponds to C1 = 1, C2 = 0. As a p-r generator, the circuit would produce the sequence defined earlier (Fig. 3.20) provided:

(a) the initial contents (values on Q1, Q2, Q3 and Q4) are not all zero, and
(b) Z1, Z2, Z3 and Z4 are set to 0.

Unfortunately, the reset mode sets Q1, Q2, Q3 and Q4 to 0.

An alternative setting for the Z inputs is given by Z1 = Z2 = Z3 = 0, Z4 = 1. Under these conditions, the LFSR generates 15 p-r patterns as before (but in a different sequence) but this time the lock-up pattern is (1110).

As a CRC signature generator, the BILBO can act in two modes - serial input or parallel input. For serial input, the sampled data are entered on Z1 with Z2, Z3 and Z4 held at 0. For parallel input, the sampled data are entered on some or all of the Z input lines. The theory of multi-input LFSRs is complex and outside the scope of this book. In practice, the properties of a multi-input LFSR are similar to those described earlier in relation to a single-input LFSR. Indeed, the single-input LFSR can be seen to be the special case of the multi-input LFSR for which the binary sequences on Z2, Z3, ... are all logic 0s.

3.5.4 Use of BILBOs in scan and self-test designs

BILBOs can be used in scan designs in a way similar to SRLs in LSSD provided care is taken to avoid the race problem. This means that for BILBOs that use simple D-type latches, rather than level-sensitive master-slave versions, the scan arrangement should be analogous to the LSSD single-latch system, as outlined in Fig. 3.29.

The BILBOs, B(1) and B(2), are independently controlled by their own clock, CLK(1) and CLK(2), and their own mode controls, C1(1)-C2(1) and C1(2)-C2(2). Similarly, the individual scan paths are kept separate as shown. For normal system operation, both pairs of mode controls are set to 1. For test operation, the sequence of events, called the Scan test sequence, is as follows.

Scan test sequence, to test N(1):

1. Set B(1) into normal system mode
2. Set B(2) into scan-path mode
3. Strobe test input vectors into B(2)
4. Apply Y(2) outputs plus PI(1) values to N(1)
5. Latch N(1) outputs into B(1)
6. Convert B(1) into scan-path mode
7. Strobe contents of B(1) out and check for correctness
8. Return to step 1 if more tests exist.

To test N(2):

1. Reverse roles of B(1) and B(2) in procedure above.

Alternatively, the following sequence, called the LFSR test sequence, exploits the LFSR properties of the BILBO.

LFSR test sequence, to test N(1):

1. Set B(1) into multi-input CRC generator mode
2. Set B(2) into p-r generator mode
3. Apply a number of p-r patterns to N(1) using the outputs of B(2), plus PI(1) if necessary
4. Accumulate N(1)'s response as a CRC signature in B(1)
5. On completion, convert B(1) into scan-path mode and strobe out the result (or provide multi-output visibility for direct comparison).

To test N(2):

1. Reverse roles of B(1) and B(2) in procedure above.

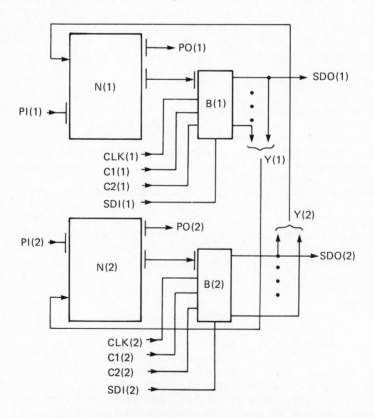

Fig. 3.29 Scan design based on BILBOs.

The major advantage of the LFSR test sequence over the Scan test sequence lies in the reduction of test time. The Scan test sequence carries a penalty of continual switching of one of the BILBOs between normal mode and scan mode. There is also a time penalty associated with the serial scan-in and serial scan-out activities. In the LFSR test sequence, the first penalty is not incurred and the second can be avoided if the parallel outputs of each BILBO are brought out to primary outputs. Other comments about the LFSR test sequence are as follows.

(a) In step 2, there should be some control over the values that can be set onto the Z inputs to B(1) in order to ensure maximal or near-maximal length to the p-r sequence. (Recall that $Z1 = Z2 = Z3 = 0$, $Z4 = 1$ ensured 15 p-r patterns. Other settings on the Z inputs will not necessarily follow suit.)

(b) Step 3 suggested 'a number' of p-r patterns, thereby begging the question - how many is enough? The constraints on the problem are the number of stages in the p-r generator and the time available. Subject to these constraints, the answer is 'as many as possible' to maximise the fault-cover. For an n-stage LFSR, the maximum number of distinct p-r patterns is $(2^n - 1)$. After this, the cycle repeats (as shown in Fig. 3.20). In general, BILBOs tend to be 8-bits long rather than 4 bits, with the feedback pattern shown in Fig. 3.30.

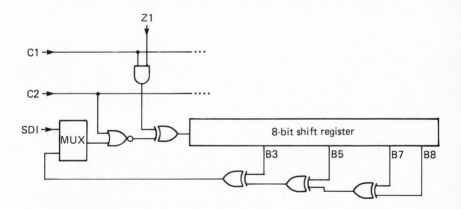

Fig.3.30 Nature of feedback structure in 8-bit BILBO.

Finally, Fig. 3.31 shows how BILBOs can be used in a self-testing circuit design based on the Signature Analysis concept.

The components of this circuit are as follows. B(1) is a BILBO configured as a p-r generator and providing the test stimuli. B(2) is another BILBO which ordinarily performs as part of the system but which, in test mode, functions as a multi-input LFSR collecting binary data and generating CRC signatures. The decoder/comparator decodes the output from B(2) (conveniently obtained at the feedback disconnect point if not directly wired) and compares it with a pre-programmed set of values. The decoder produces a single go/no go output according to whether the comparison is favourable or not. The binary up-counter is initialised to zero and provides a control on the number of clocks applied to both B(1) and B(2) when these components are in their LFSR modes.

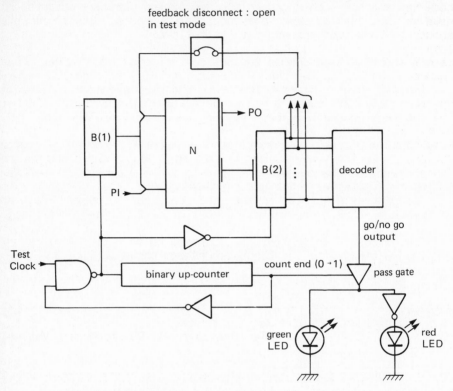

For clarity, mode control lines, normal system clocks, and preset/clear facilities have been omitted.

Fig. 3.31 Self-testing design using BILBOs.

The test operation is now simple. B(1) applies a preset number of p-r patterns to N. The output responses from N are strobed into B(2) thereby generating a series of CRC signatures. Note that the operation of B(2) is out of phase with that of B(1). The maximum speed of operation is therefore governed by the predicted worst-case propagation delay through N. On completion of the preset count sequence, the Count End signal goes high thereby inhibiting any further count and enabling the output of the decoder to influence the supply to the two on-board LEDs. A 'go' output level of 1 will turn on the green LED; a 'no go' output of 0 will turn on the red LED. Effectively, the scheme in Fig.3.31 is a more detailed form of the self-test scheme outlined in Fig.3.23 (Scheme 3).

3.6 CONCLUDING REMARKS ON SCAN DESIGN

In this chapter we have considered the concept and application of scan design followed by self test based on scan design. The principle is to incorporate facilities into the design implementation which reduce the complexity of the test-generation task to a level which makes the task soluble. There are penalties however and these have been discussed. In the end, as with all testability considerations, the final decision to incorporate a scan path is based on cost - cost of extra silicon or integrated circuits, cost of shipping devices or systems that have been inadequately tested, and so on.

The driving force for scan design has come from the large system companies with captive device manufacturing facilities. The recent announcement of the Am29818 Serial Shadow Register device has opened up scan possibilities to all logic designers irrespective of their company. This means that more and more designers will begin to put together designs with a scan-path facility. The next chapter therefore describes an effective method for generating tests for such circuits.

BIBLIOGRAPHY

SCAN DESIGN, EXCLUDING LSSD (SEE LATER SECTION)

3.1 Carter W.C., Montgomery H.C., Preiss R.J., and Reinheimer H.J., 1964, "Design of serviceability features for the IBM System 360", IBM Journal Research & Development, Vol. 8, April, pp 115-126.

3.2 Williams M.J.Y., and Angell J.B., 1973, "Enhancing testability of LSI circuits via test points and additional logic", IEEE Trans. Computers, Vol. C-22, pp 46-60.

3.3 Toth A., and Holt C., 1974, "Automated data-base-driven digital testing", IEEE Computer, Vol. 7, No. 1, pp 13-19.

3.4 Stewart J.M., 1977, "Future testing of large LSI circuit cards", Proc. IEEE. Semiconductor Test Conf., pp 6-15.

3.5 Yamada A., et al., 1977, "Automatic test generation for large digital circuit"', Proc. IEEE 14th Design Automation Conf., pp 78-83.

3.6 Yamada A., et al., 1978, "Automatic system level test generation and fault location for large digital systems", Proc. IEEE 15th Design Automation Conf., pp 347-352.

3.7 Funatsu S., Wakatsuki N., and Yamada A., 1978, "Designing digital circuits with easily testable considerations", Proc. IEEE Semiconductor Test Conf., pp 98-102.

3.8   Funatsu S., Wakatsuki N., and Yamada A., 1979, "Easily testable design of large digital circuits", NEC Journal Research and Development, No. 54, pp 49–55.

3.9   Kawai M., Funatsu S., Yamada A., 1980, "Application of shift register approach and its effective implementation", Proc. IEEE Test Conf., Paper 2.2, pp 22–25.

3.10 Trischler I., 1980, "Incomplete scan path with an automatic test generation methodology", Proc. IEEE Test Conf., Paper 7.1, pp 153–156.

3.11 Mercer M.R., Agrawal V.D., and Roman C.M., 1981, "Test generation for highly sequential scan–testable circuits through logic transformation", Proc. IEEE Test Conf., Paper 18.6, pp 561–565.

3.12 Lineback J.R., 1981, "Self–testing processor cuts costs", Electronics International, Vol. 54, No. 25, Dec. 15, pp 110–112.

RANDOM ACCESS SCAN

3.13 Audo H., 1980, "Testing VLSI with random access scan", Digest IEEE Compcon, 80CH1491–0C, pp 50–52.

3.14 Williams T.W., 1980, "Design for testability. Section 4.4: random access scan", Proc. NATO Advanced Study Institute on Computer Design Aids for VLSI Circuits, Italy.

LEVEL–SENSITIVE SCAN DESIGN

3.15 Muehldorf E.I., 1976, "Designing LSI logic for testability", Proc. IEEE Semiconductor Test Conf., pp 45–49.

3.16 Williams T.W., and Eichelberger E.B., 1977, "Random patterns within a structured sequential logic design", Proc. IEEE Semiconductor Test Conf., pp 19–26.

3.17 Williams T.W., and Eichelberger E.B., 1977, "A logic design structure for testability", Proc. IEEE 14th Design Automation Conf., pp 463–468.

3.18 Bottorf P., and Muehldorf E.I., 1977, "Impact of LSI on complex digital circuit board testing", Proc. IEEE Electro 77 Conf., pp 1–12.

3.19 Stange G.H., 1978, "A test methodology for large logic networks", Proc. IEEE 15th Design Automation Conf., pp 103–109.

3.20 Eichelberger E.B., and Williams T.W., 1978, "A logic design structure for LSI testability", Journal Design Automation and Fault Tolerant Computing, Vol. 2, pp 165–178.

3.21 Jones H.E., and Schauer R.F., 1978, "An approach to a testing system for LSI", Proc. EEC Symp. on CAD of Digital Electronic Circuits and Systems, Brussels, pp 257–274.

3.22 Stolte L.A., and Berglund N.C., 1979, "Design for testability of the IBM System 38", Proc. IEEE Test Conf., pp 29–36.

3.23 Berglund N.C., 1979, "Level–sensitive scan design tests chips, boards, systems", Electronics International, Vol. 52, No. 6, March 15, pp 108–110.

3.24 Frechette T.J., and Tanner F., 1979, "Support processor analyses errors caught by latches", Electronics International, Vol. 53, No. 23, Nov. 8, pp 116–118.

3.25 Arzoumanian Y., and Waicukauski J., 1981, "Fault diagnosis in an LSSD environment", Proc. IEEE Test Conf., Paper 4.2, pp 86–88.

3.26 Dasgupta S., Goel P., Walther R.G., and Williams T.W., 1982, "A variation of LSSD and its implications on design and test–pattern generation in VLSI", Proc. IEEE Test Conf., Paper 3.3, pp 63–66.

3.27 Komonytsky D., 1982, "LSI self–test using level–sensitive scan design and signature analysis", Proc. IEEE Test Conf., Paper 14.3, pp 414–424.

SIGNATURE ANALYSIS AND SELF–TEST, EXCLUDING
BILBO (SEE NEXT SECTION)

3.28 Gordon G., and Nadig H., 1977, "Hexadecimal signatures identify trouble spots in microprocessor systems", Electronics International, Vol. 50, No. 5, March 3, pp 89–96.

3.29 Hewlett–Packard, 1977, "A designer's guide to Signature Analysis", Application Note 222.

3.30 Hewlett–Packard, 1977, "Implementing Signature Analysis for production testing", Application Note 222–1.

3.31 Badagliacca L., and Catterton R., 1977, "Combining diagnosis and emulation yields fast fault–finding", Electronics International, Vol. 50, No. 23, Nov. 10, pp 107–110.

3.32 Nadig H., 1978, "Testing a microprocessor product using signature analysis", Proc. IEEE Semiconductor Test Conf., pp 159–169.

3.33 Neil M., and Goodner R., 1979, "Designing a serviceman's needs into microprocessor–based systems", Electronics International, Vol. 52, No. 5, March 1, pp 122–128.

3.34 Pynn C., 1979, "In–circuit tester using Signature Analysis adds digital LSI to its range", Electronics International, Vol. 52, No. 11, May 24, pp 153–157.

3.35 West D.G., 1979, "In–circuit ·emulation and signature analysis: vehicles for testing microprocessor–based products", Proc. IEEE Test Conf., pp 345–353.

3.36 Humphrey J., and Firooz K., 1980, "ATE brings speedy complete testing via signature analysis to LSI–board production", Electronic Design, Vol. 28, No. 3, Feb 1, pp 75–79.

3.37 McCluskey E.J., 1981, "Design for autonomous test", IEEE Trans. Computers, Vol. C–30, No. 11, pp 866–875.

3.38 Davidson R.P., Harrison M.L., and Wadsack R.L., 1981, "BELLMAC–32: a testable 32–bit microprocessor", Proc. IEEE Test Conf., Paper 2.2, pp 15–20.

3.39 McCluskey E.J., 1982, "Built–in verification test", Proc. IEEE Test Conf., Paper 9.2, pp 183–190.

3.40 Bardell P.H., and McAnney W.H., 1982, "Self–testing of multichip logic modules", Proc. IEEE Test Conf., Paper 9.3, pp 200–204.

3.41 Bennetts R.G., 1982, "Introduction to digital board testing", Crane–Russak (New York), Edward Arnold (London), Chap.4.

BUILT–IN LOGIC BLOCK OBSERVATION

3.42 Konemann B., Mucha J., and Zwiehoff G., 1979, "Built–In Logic Block Observation technique", Proc. IEEE Test Conf., pp 37–41.

3.43 Fasang P.P., 1980, "BIDCO, Built–In Digital Circuit Observer", Proc. IEEE Test Conf., Paper 10.2, pp 261–266.

3.44 Fasang P.P., 1982, "Circuit module implements practical self test", Electronics International, Vol. 55, No. 10, May 19, pp 164–167.

3.45 Lyman J., 1982, "Built–in test checks processor board", Electronics International, Vol. 55, No. 14, July 14, pp 51–52.

3.46 Komonytsky D., 1983, "Synthesis of techniques creates complete system self–test", Electronics International, Vol. 56, No. 5, March 10, pp 110–115.

3.47 Fasang P.P., 1983, "Microbit brings self testing on board complex microcomputers", Electronics International, Vol. 56, No. 5, March 10, pp 116–119.

4

Test generation
for scan-designed circuits

The previous chapter has presented a general-purpose design technique, called scan design, in which the requirements for test generation are reduced in complexity by including facilities to partition the circuit into its two major subsections – combinational section and stored-state devices. The overall testing strategy is to test the stored-state devices first and then to use these devices to set up input test stimuli for the combinational section. The output response of this circuit is latched back into the stored-state devices and strobed out using the scan-path mode of operation.

Generating tests for combinational circuits is, theoretically, not a problem. Classical algorithms such as Roth's D-algorithm or the Critical Path algorithm have performed successfully in practical test environments for many years. Success is generally measured in terms of percent fault-cover achievement against a pre-defined set of fault-effects such as all nodes stuck-at-1 and stuck-at-0 (s-a-1, s-a-0). The continual development of custom VLSI and gate-array techniques has renewed emphasis on efficient test-generation procedures however, and this chapter describes a fault-orientated test-generation algorithm called Path Oriented DEcision Making (PODEM), developed originally by P. Goel, to satisfy the test-generation requirements of LSSD circuits. The procedure is applicable to any combinational circuit and, in the author's opinion at least, is a worthy successor to the classical algorithms just mentioned. A description of PODEM therefore forms the core of this chapter. The bibliography at the end of the chapter lists the relevant papers on the subject. References 4.11 and 4.12 are probably the two key references.

Overall, PODEM is only part of the general process of producing tests for scan-designed circuits. In keeping with the comments made in Section 3.5.2 of the previous chapter, it is usual to precede the use of PODEM with the generation and evaluation of some pseudo-random (p-r) test inputs to obtain a reasonable level of fault coverage quickly and cheaply. The procedure developed for LSSD circuits is called RAndom Path Sensitising (RAPS) and is an example of a 'clever' pseudo-random technique. A description of RAPS follows the PODEM description although, in practice, RAPS is applied to the circuit before PODEM. The chapter concludes with a discussion of the overall strategy of testing scan-designed circuits based on the use of PODEM and RAPS.

4.1 PODEM CONVENTIONS, NOTATIONS AND CONCEPTS

PODEM is a fault-orientated test-generation procedure based on the concepts of fault-effect propagation along a sensitive path. In this respect, PODEM is similar to the D-algorithm and, indeed, makes use of the D-notation (described later) first developed for use within the D-algorithm. Where PODEM differs from the D-algorithm is in the way it determines the assignment of values to the primary inputs necessary to establish a sensitive path from the source of the fault to a primary output (PO). The assignments are determined by means of a procedure that works back from the position of the target fault, creating and satisfying nodal logic-value objectives until a logic-value assignment is made to an unassigned primary input (PI). At this point, the full effect of the new PI assignment is assessed by a simulator to determine whether all objectives have been satisfied. If the answer is 'not yet', then the procedure repeats, working from the unsatisfied objective. The method will become clearer as it is illustrated through an example. Before doing this, it is necessary to introduce and define certain notations and ideas used within PODEM.

4.1.1 Node naming

The word 'node' is used to identify a signal-carrying connection from one circuit element to another. If there is no symbolic name associated with the node, it is named according to the element and pin number of the element from which it originates, not on which it terminates. In this way, a node from an originating device that fans out to more than one terminating device is named unambiguously. The names themselves will be written 'element name . terminal number', e.g. an output terminal, pin 3, of gate G1 will be named G1.3. Figure 4.1 illustrates the naming convention applied to a simple circuit with fanout. (Note: in Fig. 4.1 and later circuit diagrams, the terminal numbers used are quite arbitrary although they tend to follow SN7400 series numbers. The reader should not attach any importance to these numbers. They are assigned simply to allow unambiguous identification.)

4.1.2 Nodal logic values.

PODEM makes use of the D-notation to indicate the fault-effect generation or transmission status of a logic node. The full set of values that a node can take is therefore as follows.

(a) A fixed value of logic 0.
(b) A fixed value of logic_1.
(c) A fixed value of D or $\bar{D}$ where $D = 1$ ($\bar{D} = 0$) corresponds to the

fault–free value on the node, and D = 0 ($\bar{D}$ = 1) defines the faulty value of the node corresponding to the node acting either as the source of the fault or as a transmitter of an earlier fault condition.
(d) An unassigned value of X, which can take any one of the 0, 1, D or $\bar{D}$ values.

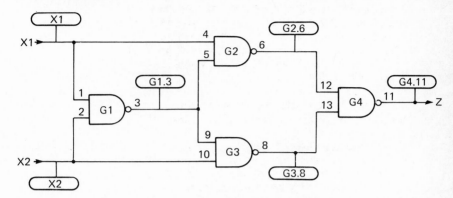

Fig. 4.1 Node–naming convention.

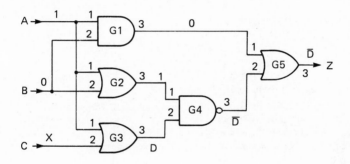

Fig. 4.2 Example of the D notation.

For those readers unfamiliar with the D notation, Fig. 4.2 shows a simple example of a circuit set up in a test condition – in this case with the fault source on G3.3 (fault–free value = 1; faulty value = 0 corresponding to the source fault G3.3 stuck–at–0) and fault–effect propagation through G4 to G5. Figure 4.2 also shows the fixed 0 and 1 values to establish the test condition for the source fault, together with an unassigned (X) value on one of the PIs.

Figure 4.3 summarises the test conditions and D–propagation conditions for each of the primitive gates together with the exclusive–OR gate. Figure 4.3(a) shows the input combinations on each gate to test either for an output s–a–1 fault ($\bar{D}$ output value) or for an output s–a–0 fault (D output value). Figure 4.3(b) defines the fixed value conditions on inputs to gates which allow further propagation of a D or $\bar{D}$ on the input side through to the output side.

86

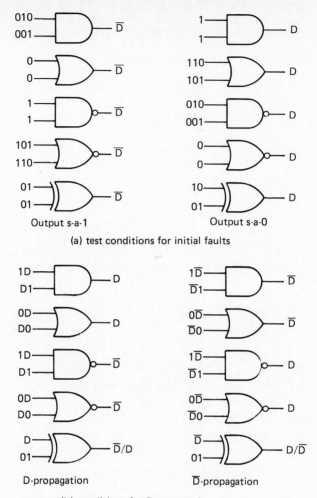

(a) test conditions for initial faults

(b) conditions for D propagation

Fig. 4.3 D notation.

4.1.3 Objectives

The detailed implementation of PODEM is based on identifying, and attempting to satisfy, certain objectives relevant to the setting of a particular node to a particular value. Examples, taken from Fig. 4.2, could be 'set G3.3 to value D' or 'set G1.3 to value 0'. Objectives such as these will be denoted OB(G3.3 = D) and OB(G1.3 = 0).

4.1.4 Objective transfer by 'easiest' or 'hardest' selection

Part of the PODEM philosophy is to attempt to satisfy a

nodal-value objective on an output node by transferring the objective to an input node based on either the 'easiest' or 'hardest' means of solution. (This operation corresponds, in concept, to the consistency operation in the D-algorithm.)

In this context, the terms easiest and hardest relate to relative controllability values of a set of input nodes and can be based on CAMELOT controllability values.

Alternatively, there is a simpler procedure, described shortly, for obtaining relative values for a set of nodes.

To illustrate the approach consider the circuit elements shown in Fig. 4.4

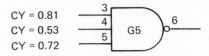

CY = 0.81
CY = 0.53
CY = 0.72

Fig. 4.4 'Easiest' and 'hardest' inputs.

The figure shows a 3-input NAND gate, G5 say, assumed to be part of a larger circuit. Controllability values are known and specified for each of the input nodes, pins 3, 4 and 5, but so far no logic values have been assigned to the node, i.e. the nodes still have the value X. Objective transfer can only take place on unassigned input nodes. Consider now the objective OB(G5.6 = 1). This objective can be satisfied by setting any one of the three inputs to a 0. It makes sense therefore to select the easiest input node on which to place a 0 – in this case, pin 3, since this node has the highest controllability. This selection transfers the output-node objective to a new nodal-value objective, OB(G5.3 = 0), and so the backtrace procedure continues in a manner to be described more fully through the examples.

Consider now the alternative objective for the value of G5.6, namely OB(G5.6 = 0). This objective can only be satisfied if all three NAND-gate inputs are set to logic 1. The backtrace procedure only pursues one objective at a time rather than create simultaneous multiple objectives. Faced with this situation, the procedure must choose which input to pursue. The most sensible choice is to select the hardest input to control, i.e. pin 4, on the grounds that if this input cannot be set to the required value, then there is no point in pursuing the easier inputs. In other words, attempt to solve the most difficult problem first rather than waste time with the easier problems. The initial objective, OB(G5.6 = 0), is then transferred to a new objective, OB(G5.4 = 1), and so the process continues.

In the event that G5.4 is actually set to the value of 1 then, inevitably, the process will come back to this gate with the original objective OB(G5.6 = 0). In this case, the selection can only be between pins 3 and 5 and the result would be OB(G5.5 = 1) since pin 5 is harder to control than pin 3. Figure 4.5 shows the 'easiest' and 'hardest' levels to satisfy 0, 1, D or $\bar{D}$ requirements on the outputs of each of the primitive gate types plus the exclusive-OR gate.

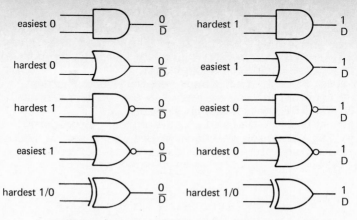

Fig. 4.5 Easiest/hardest values for primitive gates.

4.1.5 Assigning relative controllability values

The previous section discussed the use of relative controllability values to guide the selection of easiest and hardest solutions to satisfy an objective. Controllability values can either be calculated by CAMELOT or they may be determined by the following simpler but effective procedure.

STEP 1 Assign each trunk (originating) signal-carrying node an intrinsic nodal weight, IW, equal to its fanout - 1, i.e.:

$$IW \text{ (each node)} = \text{fanout} - 1.$$

STEP 2 For each circuit element whose input nodes are fully assigned, reassign the weight of each output node to be:

output node IW + sum (all input node weights)

STEP 3 Continue with Step 2 until all nodes have been reassigned. The higher the nodal weight value, the harder it is to control.

The procedure is illustrated by application to the circuit drawn earlier in Fig. 4.2.

Node	STEP 1 Intrinsic Weights	STEPS 2/3 Final Weights
A	2	2
B	1	1
C	0	0
G1.3	0	3
G2.3	0	3
G3.3	0	2
G4.3	0	5
G5.3(Z)	0	8

Notice that the order in which the nodes are processed is important. One way of ensuring that the order is correct is to use the Terminal Numbering Convention (TNC). The method is to assign a unique TNC number, starting with 1 and incrementing by 1, to each node in the circuit starting with the primary inputs. A TNC number can only be assigned to the output of a circuit element if all input nodes are already numbered, in which case, the output node will have a higher TNC number than any of the input-node numbers. The order of processing the nodes then follows the order imposed on the nodes by the TNC numbers.

For example, applying the TNC procedure to Fig. 4.2 produces the following result.

	Node	TNC No.
Group 1	A	1
	B	2
	C	3
Group 2	G1.3	4
	G2.3	5
	G3.3	6
Group 3	G4.3	7
Group 4	G5.3	8

Within each group, the assignment of TNC numbers is quite arbitrary but all nodes must be assigned before the next group can be assigned.

4.2 THE PODEM PROCEDURE

The basic approach of PODEM is to generate a test for a target fault by attempting to solve nodal-value objectives, one at a time, and in such a way as to reach a solution quickly. Fundamental to the PODEM mechanism is the backtracing technique of transferring an objective on an element output to an objective on one of the unassigned element inputs. When an element input becomes a primary input, the objective is satisfied by making the appropriate assignment to the value of the primary input. This is called a PI-assignment point and it triggers the call to a 5-value $(0,1,X,D,\bar{D})$ zero-delay simulator to determine the full effect of the PI assignment. The requirement to produce a test may or may not be satisfied by this new PI assignment. If it is, the job is done. If it is not, then PODEM is re-entered at the point determined by the furthest position of the propagating D or $\bar{D}$, and a further attempt is made to satisfy the requirement for a test by specifying an objective based on furthering the fault-effect propagation (a process called D-drive). PODEM continues in this way until either a test is found or it is proved that no test exists. This feature of the procedure, which makes it an algorithm, will be demonstrated in one of the later examples.

What we will do now is to demonstrate the detailed mechanisms of PODEM by specific examples applied to the equivalent gate-level circuit of the SN7480 gated full adder, shown in Fig. 4.6. Note that, as before, the numbers assigned to gates within this circuit are arbitrary.

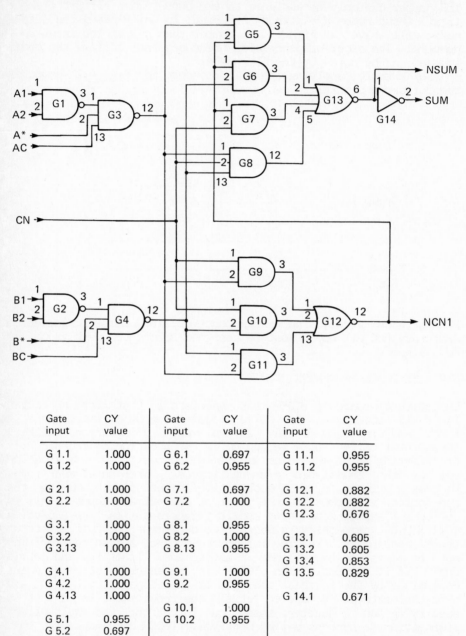

Gate input	CY value	Gate input	CY value	Gate input	CY value
G 1.1	1.000	G 6.1	0.697	G 11.1	0.955
G 1.2	1.000	G 6.2	0.955	G 11.2	0.955
G 2.1	1.000	G 7.1	0.697	G 12.1	0.882
G 2.2	1.000	G 7.2	1.000	G 12.2	0.882
				G 12.3	0.676
G 3.1	1.000	G 8.1	0.955		
G 3.2	1.000	G 8.2	1.000	G 13.1	0.605
G 3.13	1.000	G 8.13	0.955	G 13.2	0.605
				G 13.4	0.853
G 4.1	1.000	G 9.1	1.000	G 13.5	0.829
G 4.2	1.000	G 9.2	0.955		
G 4.13	1.000			G 14.1	0.671
		G 10.1	1.000		
G 5.1	0.955	G 10.2	0.955		
G 5.2	0.697				

Fig. 4.6 SN7480 gated full adder

The reader who does not wish to pursue the detail of the algorithm at this stage may proceed directly to Section 4.3 which discusses the pseudo-random test generator called RAPS. Figure 4.8 provides a top-level flowchart of the main PODEM activities.

4.2.1 Example 1: Basic PODEM

This first example demonstrates a straightfoward application of PODEM and is fully detailed to the point at which the test is derived. The reader is recommended to follow the procedure by marking up the nodal values on a photocopy of Fig. 4.6.

CAMELOT controllability values are listed in Fig. 4.6 for each node. The closer the value to 1, the easier the node is to control.

INITIALISE all nodal values to unassigned value X

SELECT target fault: G8.12 s-a-1 (Arbitrarily selected;
                             defines start D objective)

START:
OB(G8.12 = $\overline{D}$)

BACKTRACE:
Set easiest G8 X-input to 0: G8.2 (= CN)
OB(CN = 0)

ASSIGN PI(CN) = 0   PI ASSIGNMENT

SIMULATE effect of new PI assignment
CN = 0
G7.3 = 0
G8.12 = $\overline{D}$
G9.3 = 0
G10.3 = 0

QN: Is the D objective satisfied?
Yes
QN: Is a D propagated to a PO?
No
Determine shortest X-path route to a PO from current D point
Shortest route is to NSUM. Arbitrary choice between
equal-length shortest paths through G10 or G11.
Use D-drive to establish next D objective: OB(G13.6 = D)

CONTINUE D-DRIVE
OB(G13.6 = D)

BACKTRACE:
Set hardest G13 X-input to 0: G13.2, arbitrarily (= G6.3)
OB(G6.3 = 0)
Set easiest G6 X-input to 0: G6.2 (= G4.12)
OB(G4.12 = 0)

Set hardest G4 X–input to 1: G4.1, arbitrarily (= G2.3)
OB(G2.3 = 1)
Set easiest G2 X–input to 0: G2.1, arbitrarily (= B1)
OB(B1 = 0)

ASSIGN PI(B1) = 0   PI ASSIGNMENT

SIMULATE effect of new PI assignment
B1 = 0
G2.3 = 1

QN: Is the D objective satisfied?
No
Continue with the D objective

OB(G13.6 = D)

BACKTRACE:
Set hardest G13 X–input to 0: G13.1 arbitrarily (= G5.3)
OB(G5.3 = 0)
Set easiest G5 X–input to 0: G5.1 (= G3.12)
OB(G3.12 = 0)
Set hardest G3 X–input to 1: G3.2, arbitrarily (= A*)
OB(A* = 1)

ASSIGN PI(A*) = 1   PI ASSIGNMENT

SIMULATE effect of new PI assignment
A* = 1

QN: Is the D objective satisfied?
No
Continue with the D objective

OB(G13.6 = D)

BACKTRACE:
Set hardest G13 X–input to 0: G13.1, arbitrarily (= G5.3)
OB(G5.3 = 0)
Set easiest G5 X–input to 0: G5.1 (= G3.12)
OB(G3.12 = 0)
Set hardest G3 X–input to 1: G3.13, arbitrarily (= AC)
OB(AC = 1)

ASSIGN PI(AC) = 1   PI ASSIGNMENT

SIMULATE effect of new PI assignment
AC = 1

QN: Is the D objective satisfied?
No
Continue with the D objective

OB(G13.6 = D)

BACKTRACE:
Set hardest G13 X-input to 0: G13.2, arbitrarily (= G6.3)
OB(G6.3 = 0)
Set easiest G6 X-input to 0: G6.2 (= G4.12)
OB(G4.12 = 0)
Set hardest G4 X-input to 1: G4.2, arbitrarily (= B*)
OB(B* = 1)

ASSIGN PI(B*) = 1   PI ASSIGNMENT

SIMULATE effect of new PI assignment
B* = 1

QN: Is the D objectve satisfied?
No
Continue with the D objective

OB(G13.6 = D)

BACKTRACE:
Set hardest G13 X-input to 0: G13.1, arbitrarily (= G5.3)
OB(G5.3 = 0)
Set easiest G5 X-input to 0: G5.1 (= G3.12)
OB(G3.12 = 0)
Set hardest G3 X-input to 1: G3.1 (= G1.3)
OB(G1.3 = 1)
Set easiest G1 X-input to 0: G1.2, arbitrarily (= A2)
OB(A2 = 0)

ASSIGN PI(A2) = 0   PI ASSIGNMENT

SIMULATE effect of new PI assignment
A2 = 0
G1.3 = 1
G3.12 = 0
G5.3 = 0
G8.12 = $\bar{D}$, as before
G9.3 = 0, as before
G11.3 = 0
G12.12 = 1
NCN1 = 1   PO ASSIGNMENT

QN: Is the D objective satisfied?
No
Continue with the D objective

OB(G13.6 = D)

BACKTRACE:
Set hardest G13 X-input to 0: G13.2 (= G6.3)
OB(G6.3 = 0)
Set easiest G6 X-input to 0: G6.2 (= G4.12)
OB(G4.12 = 0)
Set hardest G4 X-input to 1: G4.13 (= BC)

94

OB(BC = 1)

ASSIGN PI(BC) = 1   PI ASSIGNMENT

SIMULATE effect of new PI assignment
BC = 1
G4.12 = 0
G6.3 = 0
G8.12 = $\bar{D}$, as before
G10.3 = 0, as before
G11.3 = 0, as before
G13.6 = D
NSUM = D   PO ASSIGNMENT
G14.2 = $\bar{D}$
SUM = $\bar{D}$ PO ASSIGNMENT

QN: Is the D objective satisfied?
Yes
QN: Is a D propagated to a PO?
Yes

TEST is found for target fault:

A1	A2	A*	AC	CN	B1	B2	B*	BC	NSUM	SUM	NCN1
X	0	1	1	0	0	X	1	1	1(D)	0($\bar{D}$)	1

EXIT

Note the following points from this example.

(a) The simulation process checks for conflicting values on assigned nodes. In this example, there were no conflicting values but this is not always so as will be demonstrated in a later example.
(b) The final test has two PIs still unassigned, A1 and B2. Their values can either be assigned with random values of 0 or 1 or hey can be left unassigned to allow for the possibility of subsequent merging of two or more tests into a single composite test. This merging process is termed compaction and is discussed in a later section.

4.2.2 Example 2: multi–path D–drive

In essence, the backtrace procedure is very simple. The transfer of objectives is easily computed and, for a combinational circuit, must always terminate at a primary input. This next example, still based on Fig. 4.6, illustrates a strategy for handling D–drive from a node with fanout plus multi–path choices. Some of the backtracing detail is now omitted and a shorter descriptive form of objective transfers is used.

INITIALISE all nodal values to unassigned value X

SELECT target fault: G4.12 s–a–0

```
START:
OB(G4.12 = D)

BACKTRACE:
OB(G4.12 = D):- G4.2, arbitrarily (= B*) = 0
OB(B* = 0)

ASSIGN PI(B*) = 0                                    PI ASSIGNMENT

SIMULATE:
B* = 0
G4.12 = D

QN: Is the D objective satisfied?
Yes
QN: Is a D propagated to a PO?
No
Determine shortest X-path route to a PO.
Shortest route is to NCN1.  Arbitrary choice between
equal-length shortest paths through G10 or G11
Use D-drive to establish next objective: OB(G10.3 = D)

CONTINUE D-DRIVE
OB(G10.3 = D)

BACKTRACE:
OB(G10.3 = D):- G10.1 (= CN) = 1
OB(CN = 1)                                           PI ASSIGNMENT

SIMULATE:
CN = 1
G10.3 = D

QN: Is the D objective satisfied?
Yes
QN: Is a D propagated to a PO?
No
Determine shortest X-path to a PO. (Still NCN1)
Use D-drive for next objective: OB(G12.12 = D̄)

CONTINUE D-DRIVE
OB(G12.12 = D̄)

BACKTRACE:
OB(G12.12 = D̄):- G12.13 (= G11.3) = 0
OB(G11.3 = 0):- G11.2  (= G3.12) = 0
OB(G3.12 = 0):-
              :
              :
OB(A1 = 0)                                           PI ASSIGNMENT
    :
OB(A* = 1)                                           PI ASSIGNMENT
    :
OB(AC = 1)                                           PI ASSIGNMENT
```

96

SIMULATE:
A1 = 0
A* = 1
AC = 1
G1.3 = 1
G3.12 = 0
G5.3 = 0
G8.12 = 0
G9.3 = 0
G11.3 = 0
G12.12 = $\bar{D}$
NCN1 = $\bar{D}$ PO ASSIGNMENT
G6.3 = 0 (Negative reconvergence: D.$\bar{D}$ = 0)
G7.3 = $\bar{D}$
G13.6 = D
NSUM = D   PO ASSIGNMENT
G14.2 = $\bar{D}$
SUM = $\bar{D}$ PO ASSIGNMENT

The sequence of questions will now exit with the following test:

A1	A2	A*	AC	CN	B1	B2	B*	BC	NSUM	SUM	NCN1
0	X	1	1	1	X	X	0	X	1(D)	0($\bar{D}$)	0($\bar{D}$)

Comments on this example are as follows.

(a) The routine for finding X-paths looks for the shortest X-path from the D-point to any PO. If there is a choice, the selection is arbitrary.
(b) Note the example of negative fault-effect reconvergence on G6.

4.2.3 Example 3: PI REMAKE

The next example is based on Fig. 4.7 and illustrates the algorithmic property of PODEM. This property is based on a process, called PI-REMAKE, whereby the value of each assigned PI is systematically inverted as conflict arises in order to determine whether there exists any set of input values which tests for the target fault. Ultimately, all sets of values are tried.

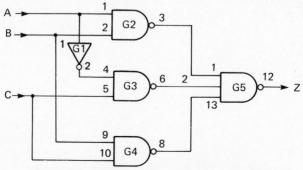

Fig. 4.7 Circuit for PODEM example 3.

INITIALISE all nodal values to unassigned value X

SELECT target fault: G4.8 s–a–1

START:
OB(G4.8 = $\bar{D}$)

BACKTRACE:
OB(G4.8 = $\bar{D}$):– G4.9, arbitrarily (= B) = 1
OB(B = 1)

ASSIGN PI(B) = 1   PI ASSIGNMENT No.1

SIMULATE:
B = 1

Question sequence returns to OB(G4.8 = $\bar{D}$)

BACKTRACE:
OB(G4.8 = $\bar{D}$): G4.10 (= C) = 1
OB(C = 1)

ASSIGN PI(C) = 1   PI ASSIGNMENT No.2

SIMULATE:
C = 1
G4.8 = $\bar{D}$

Question sequence now produces OB(G5.12 = D)

BACKTRACE:
OB(G5.12 = D):– G5.1, arbitrarily (= G2.3) = 1
OB(G2.3 = 1) :– G2.1 (= A) = 0
OB(A = 0)

ASSIGN PI(A) = 0   PI ASSIGNMENT No.3

SIMULATE:
A = 0
G2.3 = 1
G1.2 = 1
G3.6 = 0
G5.12 = 1   CONFLICT with required value of D

APPLY PI–REMAKE:
Re–assign the last unmarked assigned PI to its complemented
value. Mark as re–assigned (R) and re–simulate

RE–ASSIGN PI(A) = 1   PI ASSIGNMENT No.3/R

SIMULATE:
A = 1
G1.2 = 0
G3.6 = 1

G2.3 = 0
G5.12 = 1   CONFLICT with required value of D

APPLY PI-REMAKE:
Re-assign the last unmarked assigned PI to its complemented
value.
Mark as re-assigned (R) and re-simulate

RE-ASSIGN PI(C) = 0   PI ASSIGNMENT No.2/R

SIMULATE
        .
        .
        .
     (etc)

For this example the procedure will eventually back-up all the
way to the first PI-assignment and still fail to find a test for
the target fault (because the fault lies on a logically redundant
node).   As can be seen, the PI-REMAKE procedure is simply a
mechanism for trying all possible combinations: this is what makes
PODEM an algorithm.

In practice, extensive PI-REMAKE occurs only for circuits with
complex fanout and reconverging features or for circuits with
logically redundant (i.e.untestable) nodes.   This observation
leads to an alternative strategy whereby the number of PI-REMAKES
is limited (to n say, where n is the number of PIs).   This
approach reduces the amount of potentially wasted computation time
but removes the algorithmic property of PODEM.

4.2.4 Summary of PODEM algorithm

This section concludes the discussion of PODEM.4.8 Figure 4.8
summarises the top-level flow of activities in PODEM.

Figure 4.9 presents a circuit on which the reader might like
to exercise his or her understanding of PODEM. The circuit, known
as the Schneider Counterexample, has a complex but symmetrical
structure. A particularly interesting fault for which to generate
a test is G2.4 s-a-0.

4.3   THE RAPS PROCEDURE

In the previous chapter, we discussed the value of pseudo-random
input patterns to combinational logic circuits in terms of their
ability to provide substantial fault-cover quickly.   The RAndom
Path Sensitising (RAPS) procedure is a variant of PODEM which
seeks to produce an assignment of values to the circuit PIs which
have a better-than-average chance of covering uncovered fault
conditions. RAPS does this by deliberately attempting to maximise
the number of sensitive paths through to the circuit POs. For
circuits with a high degree of fanout and fanin, RAPS has been
found to be more effective (higher fault-cover per test) than
pseudo-random techniques based on, for example, a linear feedback
shift register.

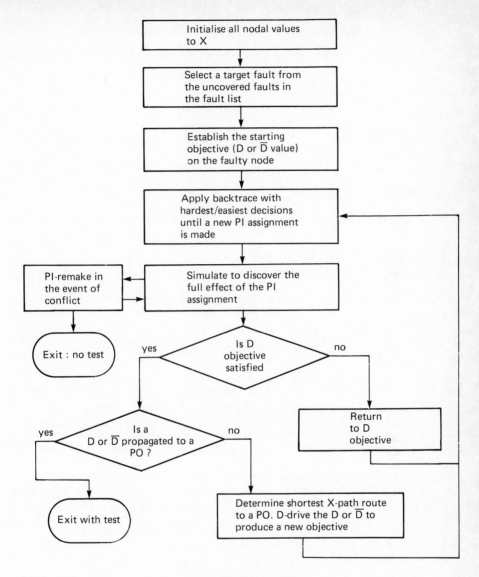

Fig.4.8 PODEM flowchart.

There are three main differences between PODEM and RAPS.

(a) The starting point for PODEM is 'find a test for the following specified fault condition'. The starting point for RAPS is 'select any PO at random and assign it a random value, 0 or 1'. In other words, PODEM is fault-orientated whereas RAPS is fault-independent.

(b) RAPS uses the same backtracing process as PODEM in its attempt to create PI input values that satisfy the PO assignment, but the concept of 'hardest/easiest' is replaced with 'random'.

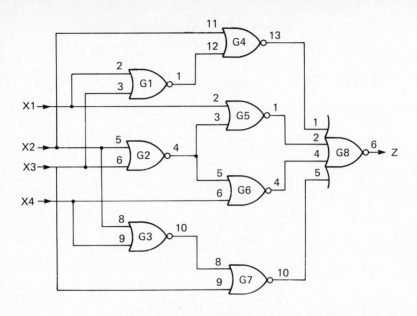

Fig. 4.9   Schneider counterexample circuit.

(c) Each pass through RAPS is independent of any other pass.   The only information retained during a RAPS pass is the PI assignments and derived internal values.side Even the assignment to the chosen PO is not recorded.   This means that RAPS can produce a set of PI assignments which sets the chosen PO to the complement of its initial value.   This does not matter.   The objective of RAPS is to provide high fault cover by maximising the number of sensitive paths.   Starting from a chosen PO with an assignment objective is a convenient start point for this overall objective.

As with PODEM, we will consolidate our understanding of RAPS by working through an example based on Fig. 4.6.   We will use the shorthand notation:

$$OB(G3.12 = 0) :- G3.2/R (= A^*) = 1$$

to mean that the objective of setting a gate output G3.12 to 0 is partially achieved by selecting, at random, one of the X-value inputs to G3, assumed to be G3.2 in this case, and setting this input to 1.   In this example, G3.2 is connected to A*.   The letter 'R' after G3.2 denotes that the choice of G3.2 was random from amongst the inputs to G3 still at X.   Note that the value of 1 selected for G3.2 is definitely not a random choice.   It is determined by the type of gate.

4.3.1 Example 4: RAPS

INITIALISE all nodal values to unassigned value X

SELECT target PO at random: PO(NCN1)
SELECT initial assignment at random: 1

START:
OB(NCN1 = 1):- G12.13/R (= G11.3) = 0
OB(G11.3 = 0):- G11.2/R (= G3.12) = 0
OB(G3.12 = 0):- G3.1/R   (= G1.3) = 1
OB(G1.3 = 1):- G1.2/R   (= A2)   = 0
OB(A2 = 0)

ASSIGN PI(A2) = 0   PI ASSIGNMENT

SIMULATE:
A2 = 0
G1.3 = 1

QN: Is the selected PO assigned?
No
Continue with PO objective

START:
OB(NCN1 = 1):- G12.2/R (= G10.3) = 0
OB(G10.3 = 0):- G10.1/R (= CN) = 0
OB(CN = 0)

ASSIGN PI(CN) = 0   PI ASSIGNMENT

SIMULATE:
CN = 0
G7.3 = 0
G8.12 = 0
G9.3 = 0
G10.3 = 0

QN: Is the selected PO assigned?
No
Continue with PO objective

START:
OB(NCN1 = 1):- G12.13 (= G11.3) = 0   (Single choice)
OB(G11.3 = 0):- G11.2/R (= G3.12) = 0
OB(G3.12 = 0):- G3.13/R (= AC) = 1

ASSIGN PI(AC) = 1   PI ASSIGNMENT

SIMULATE:
AC = 1

QN: Is the selected PO assigned?
No

Continue with PO objective

START:
OB(NCN1 = 1):– G12.13 (= G11.3) = 0   (Single choice)
OB(G11.3 = 0):– G11.2/R (= G3.12) = 0
OB(G3.12 = 0):– G3.2 (= A*) = 1   (Single choice)
OB(A* = 1)

ASSIGN PI(A*) = 1   PI ASSIGNMENT

SIMULATE:
A* = 1
G3.12 = 0
G5.3 = 0
G8.12 = 0   (As before)
G9.3 = 0   (As before)
G11.3 = 0
G12.12 = 1
NCN1 = 1

        PO ASSIGNMENT

QN: Is the selected PO assigned?
Yes

Comment: the strategy now is to select another PO whose value, as yet, is still unassigned and to continue with the backtracing process. Note that a fair number of internal node values are now already assigned.

QN: Are there any POs still at X?
Yes

SELECT new target PO at random: NSUM
SELECT initial assignment at random: 0

START:
OB(NSUM = 0):– G13.2 (= G6.3) = 1   (Single choice)
OB(G6.3 = 1):– G6.2 (= G4.12) = 1   (Single choice)
OB(G4.12 = 1):– G4.2/R (= B*) = 0

ASSIGN PI(B*) = 0   PI ASSIGNMENT

SIMULATE:
B* = 0
G4.12 = 1
G6.3 = 1
G13.6 = 0
NSUM = 0   PO ASSIGNMENT
G14.2 = 1
SUM = 1   PO ASSIGNMENT

QN: Is the selected PO assigned?
Yes
QN: Are there any POs still at X?
No

Comment: the strategy now is to test for any remaining PIs still at X. If the answer is 'no' then the pass is complete. If the answer is 'yes' then a search is made for a gate whose output is assigned but which has one or more inputs unassigned. Once identified, the objective is to set each unassigned input to the value that improves the chances of creating extra sensitive paths. For AND and NAND gates, this value is 1. For OR and NOR gates, the value is 0. For exclusive-OR gates, the value is a random choice from 0 or 1.

Continuing with the example:

QN: Are there any PIs still at X?
Yes
QN: Can a gate be found whose output is assigned but with an input still at X?
Yes: gate G4 of type NAND. G4.1 and G4.13 still at X.

START:
OB(G4.1 = 1)
OB(G2.3 = 1):- G2.1/R (= B1) = 0
OB(B1 = 0)

ASSIGN PI(B1) = 0   PI ASSIGNMENT

SIMULATE:
B1 = 0
G2.3 = 1

The sequence of questions will now return:

OB(G4.13 = 1) leading to PI(BC) = 1
OB(G2.2 = 1) leading to PI(B2) = 1
OB(G1.1 = 1) leading to PI(A1) = 1
All PIs are now assigned

Final test given by:

A1	A2	A*	AC	CN	B1	B2	B*	BC	NSUM	SUM	NCN1
1	0	1	1	0	0	1	0	1	0	1	1

with full fault-cover determined by the fault simulator to be:

A2 s-a-1, A* s-a-0, AC s-a-0, CN s-a-1, B* s-a-1 G1.3 s-a-0,
G3.12 s-a-1, G4.12 s-a-0, G9.3 s-a-1, G10.3 s-a-1,
G11.3 s-a-1,G12.12 s-a-0, G6.3 s-a-0, G13.6 s-a-1,
G14.2 s-a-0

This list represents 15 out of 46 PI and gate output s-a-1, s-a-0 faults, i.e. 33% of the single stuck-at nodal faults covered.

4.3.2 Summary of RAPS procedure

RAPS is not an algorithm insofar as it does not guarantee to find a solution to a problem. It is merely a procedure to produce useful assignments to the PIs. Figure 4.10 summarises the top-level flow of activities through RAPS.

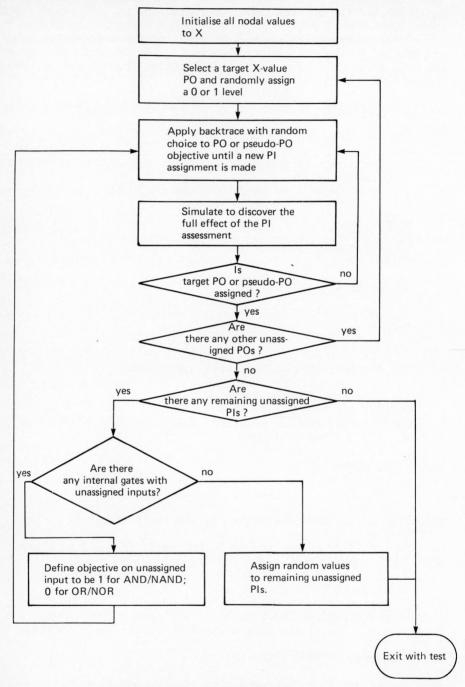

Fig. 4.10 RAPS flowchart.

4.4 RAPS/PODEM STRATEGIES

In general, RAPS is used before PODEM to provide a reasonably high level of fault–coverage before passing over to PODEM. Figure 4.11 shows one particular strategy based on the use of a fault simulator to assess individual and cumulative fault cover. Points about Fig. 4.11 are as follows.

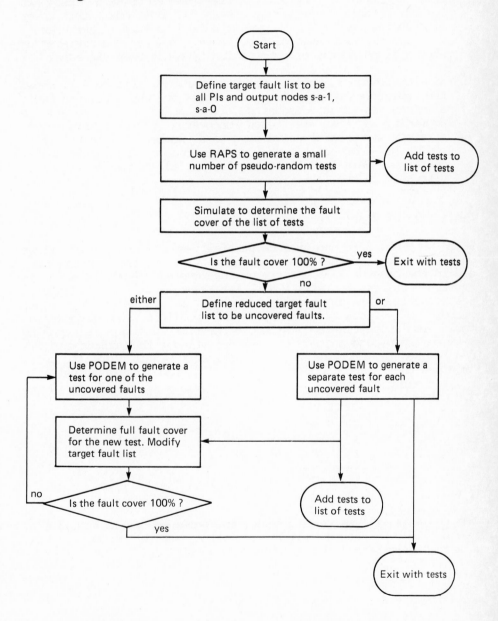

Fig. 4.11 Strategy based on RAPS and PODEM

(a) RAPS is used to generate 'a small number of p-r tests'. In practice, this number is very small, i.e. between 5 and 15, irrespective of the size of the circuit. This is because the best return in terms of fault-cover comes from the early tests, e.g. the single test derived by RAPS in Section 4.3.1 succeeded in covering one-third of the faults in the circuit. An alternative to generating a fixed number of RAPS tests is to use a fault simulator to assess the additional fault cover of each new test as it is generated and to exit either when a certain minimum level of fault cover (70% say) is achieved or when a fixed amount of computation time has been exceeded.

(b) Similar comments apply to the use of PODEM. Figure 4.11 shows two possible strategies: one based on the use of a fault simulator as just discussed; the other based purely on providing a test for each uncovered fault. This second strategy may possibly give rise to a number of tests which could be combined together to help reduce the overall number of tests. Merging of tests is called compaction and is discussed in the following sections.

4.4.1 Use of static test compaction

Both PODEM and RAPS can produce partially-specified sets of values on PIs which, in the case of PODEM satisfy the test requirement, and in the case of RAPS satisfy an intermediate objective such as 'assign values to all POs' (see Example 1 and Example 4 respectively). If an overall objective is to keep the final number of tests as small as possible (to reduce the fault-simulation time and to reduce the time penalty of cycling between scan-path and normal-path operation), then unassigned PIs can be used to merge different tests into a single composite test, provided there is no conflict between individual assigned values.
    For example, consider the two PI sets, Sets 1 and 2, on inputs A,B,C,D:

A	B	C	D	
1	1	0	X	Set 1
X	1	X	0	Set 2
1	1	0	0	Combined set

The two sets will merge to produce the combined set (A = 1, B = 1, C = 0, D = 0) as a single composite set of input values. Alternatively Sets 1 and 3 below:

A	B	C	D	
1	1	0	X	Set 1
X	0	X	0	Set 3

will not combine because of the conflicting values required on B.
    Merging tests along these lines is called static compaction

---

and this can be built into the way in which both RAPS and PODEM are used as shown in Figs. 4.12 and 4.13 respectively.

In Fig. 4.12, the exit from RAPS is shown to be controlled by the number-of-tests criterion. There is little point in considering the alternative strategy of using a fault simulator to determine cumulative fault-cover, because each RAPS test is now incompletely specified. This would result in an inconclusive output from the fault simulator.

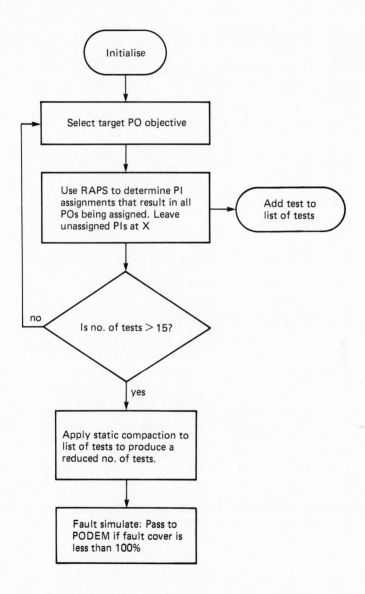

Fig. 4.12 RAPS plus static compaction.

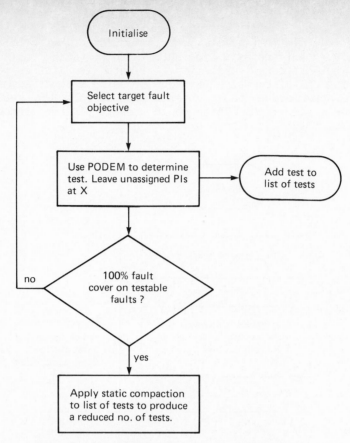

Fig. 4.13 PODEM plus static compaction.

In Fig. 4.13, the strategy in respect of the use of a fault simulator is different. Assuming the initial number of uncovered faults to be reasonably small (around 30% of the full target fault list, say), then it is probably quicker to use PODEM separately for each fault than to fault simulate after each PODEM run. This is because the increase in additional fault–coverage per new test decreases as the cumulative fault–cover given by all the tests gets closer to the 100% mark. This can be seen more clearly by looking at the classic fault–cover curve shown in Fig. 4.14.

When 100% fault–cover is achieved, static compaction can be performed on the set of tests. Be warned however; static compaction can become very time–consuming if it is based on an exhaustive search for the minimal solution. In practice, a simple search–and–combine procedure is used which does not guarantee minimality but which is completed within a reasonable computation time.

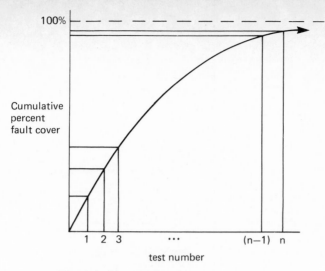

Fig. 4.14 Fault–cover curve.

4.4.2 Use of dynamic test compaction

Static test compaction is a means of reducing the total number of tests by merging separate tests together after they have all been generated. An alternative technique, used only in PODEM, is to try to maximise the additional fault–cover per test as each test is generated. The method is called dynamic test compaction and is outlined in Fig. 4.15.

As PODEM succeeds in generating a test, a count is made of the remaining number of unassigned PIs. If this number represents a reasonable percentage of the total number of PIs (15% or more is suggested as reasonable), a search is made to discover whether there are any further uncovered faults which could also be tested by additional assignments to the unassigned PIs. The search is based on finding an X–value node which represents an uncovered fault and which also lies on an X–path to an unassigned PO. If the search is successful, a new fault objective based on this secondary target–fault is specified and PODEM invoked to attempt further assignments to generate a test. This attempt may or may not be successful because certain PI and internal nodal values have already been assigned by the earlier PODEM run. If the attempt is successful, control moves back to the question concerning the remaining number of unassigned PIs. If the attempt is unsuccessful, a number of options are possible. The first is to return to the secondary target fault selection point and see whether an alternative secondary target fault exists. The second is to abandon the attempt and exit after assigning random values to unassigned PIs. One way of deciding whether to try again is to monitor the amount of time spent in the secondary target fault loop and compare with the time spent in the primary section (initial target fault). When the secondary loop time exceeds, say, 50% of the primary loop time, then is the time to abandon the secondary loop.

110

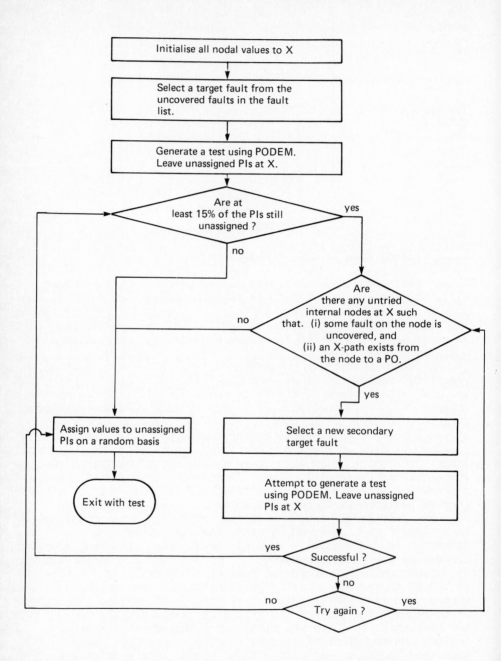

Fig. 4.15 Dynamic test compaction in PODEM.

The advantage of dynamic test compaction is that the additional fault-cover per test is maximised thereby ensuring quicker test-generation time and reduced fault-simulation time. The disadvantage is the potential amount of wasted computation during the secondary target fault loop. A compromise solution could be to attempt only one secondary target fault and abandon the search for a test as soon as conflict is identified.

4.5   COMMENT ON FAULT SIMULATION

The discussion of the PODEM and RAPS procedures has assumed the existence of an efficient means of fault simulation. There are several techniques for simulating the behaviour of logic circuits in the presence of faults but a detailed discussion of the techniques is outside the scope of this book. Nevertheless, the fault simulator is an important and integral part of the overall PODEM test-generation process and this section comments briefly on the technique that has been advocated by the developers of PODEM.

The basic requirement is for an efficient means of simulating the behaviour of a combinational logic circuit with a given input pattern to determine which faults, if any, are detected at the circuit's primary outputs. 'Efficient' means fast and economical of computer cpu time. Techniques for fault simulation fall into three main categories: serial, parallel and concurrent. Serial fault simulation determines the behaviour of the circuit in the presence of a fault by considering only a single fault per pass of the simulation. The parallel and concurrent techniques simulate multiple versions of the circuit per pass, where each version corresponds to a particular simulated fault condition. Parallel and concurrent techniques are more sophisticated than the serial technique and have been developed to serve largely as general-purpose simulators handling a wide range of circuit types – combinational and sequential. The serial technique however, has been found to be distinctly advantageous for combinational circuits where the simplicity of the circuit is matched by the simplicity of the simulator. The technique, now referred to as Single Fault Propagation (SFP), consists of the following main steps.

STEP 1   Set up the Target Fault List (TFL), Detected Fault List (DFL) and Set Of Tests (SOT). The TFL contains all single s-a-1, s-a-0 faults on primary inputs and gate output nodes. The DFL is initially empty. The SOT contains those tests specified by RAPS or generated by PODEM.

STEP 2   Apply the first test pattern from the SOT to the circuit and determine the correct nodal values throughout the circuit.

STEP 3   For each gate with an output value of 1 (0), simulate the effect of the output s-a-0 (s-a-1) provided such faults have not been detected by a previous test, i.e. provided the fault is still in the TFL, not the DFL.

STEP 4   If the effect of the gate output fault is propagated to an observable output node (circuit primary output or scan latch input), the fault is detected by this test: otherwise the fault is not detected. Delete the detected faults from the TFL and add to the DFL. Mark the test in the SOT as having been evaluated.

STEP 5   Repeat Steps 2–4 for each unmarked test in the SOT.

The overall speed of carrying out the fault simulation can be improved by the following techniques.

(a) In Step 1, the TFL can be reduced by an a priori analysis of the equivalent fault groups within the circuit. Two or more faults are equivalent if they are all detected by the same set of tests. Providing a test for any one of the faults will automatically provide a test for all the other equivalent faults. As an example, consider a 3–input NAND gate. The only test for any of the three inputs s–a–0 (single or multiple) is the all–1s input producing an expected correct output value of 0. This test happens to be the only test that sets the output low and hence tests for the output s–a–1. Effectively therefore the three input s–a–0 faults and the output s–a–1 fault are equivalent. As a result, all but one of the faults can be removed from the TFL. (From the point of view of test generation, it does not usually matter which fault is retained in the TFL. From the point of view of fault simulation it is usually more convenient to retain the gate output fault.)

The concepts of fault equivalence form the base for a fault–collapsing procedure in which local equivalent faults (faults related to the same device as in the NAND gate example above) are matched together to form larger global groups. For example, any NOR gate input s–a–1 is equivalent to the NOR–gate output s–a–0. Consequently, for a NAND gate driving a NOR gate, the following group of faults would all be equivalent:

        any NAND–gate input s–a–0
        the NAND–gate output s–a–1
        any NOR–gate input s–a–1
        the NOR–gate output s–a–0

This single set of equivalent faults amounts to 50% of all the single s–a–1, s–a–0 faults in the two–gate circuit. A test covering any one of the faults will automatically cover all the other faults.

The reduction in the TFL by the removal of equivalent faults usually results in an overall reduction of around 50% of the initial number of faults. This reduction is significant in terms of the time saved in repeated fault simulation in Step 3.

(b) Determination of the nodal values in Steps 2 and 3 can be achieved in one pass provided the sequence for evaluating the output of each gate is correct, i.e. the output value of a gate is only evaluated when all input values are known. Use of the Terminal Numbering Convention (TNC), described earlier in Section 4.1.5, ensures a correct evaluation sequence.

(c) The selection of a fault in Step 3 is governed by two questions: is the fault still in the TFL and is the fault potentially observable for this test? A fault on a particular node is potentially observable if conversion of the fixed logic value on the node to the unknown X value results in an X value at one or more of the observable output nodes. If an X value is not propagated to an observable output node, then the fault on that node and any other X-valued node is non-observable. The only faults worth simulating are those that are potentially observable.

It is claimed that, in practice, a preliminary X-propagation phase prior to the full fault simulation is very effective in reducing the total fault simulation time. The time lost in carrying out the X-propagation process is more than compensated for by the time saved in not carrying out unnecessary fault simulation.

4.6 CONCLUDING REMARKS ON PODEM/RAPS

The test generation task for a scan-designed circuit is significant only for the combinational section of the circuit. This problem has received considerable attention from test programmers and research workers alike and it seems invidious to select just one algorithm, PODEM, and devote a whole chapter to its description. In reality however, most practical test-generation algorithms are but variations on a basic theme of path sensitisation. In this respect, PODEM is no different from many other algorithms. Where it does differ however is in its practical efficiency coupled with simplicity of approach. Not only that, the algorithm was developed primarily for scan-design circuits (in this case, LSSD) and it has proved remarkably effective for circuits containing upto 50,000 gates - see reference 4.12.

For these reasons therefore, the algorithm was selected for detailed description in the book. But, there is nothing to prevent the reader from making use of other well-known algorithms, such as the D-algorithm or critical-path algorithm, to solve the test-generation requirement for a scan circuit. If other test-generation algorithms already exist in programmed form and perform satisfactorily, they can be used. If this is not the case, the PODEM/RAPS procedures that have been described in this chapter are recommended.

BIBLIOGRAPHY

4.1 Snethen T.J., 1977, "Simulator-oriented fault test generator", Proc. 14th IEEE Design Automation Conf., pp 88-93.

4.2 Bottorff P.S., et al., 1977, "Test generation for large logic networks", Proc. 14th IEEE Design Automation Conf., pp 479-485.

4.3 Goel P., 1978, "Dynamic subsumation of test patterns for LSSD systems", IBM Technical Disclosure Bulletin, Vol. 21, No. 7, pp 2782-2784.

4.4 Goel P., 1978, "Method for simultaneous generation of multiple tests for LSSD logic circuits", IBM Technical Disclosure Bulletin, Vol. 21, No. 7, pp 2785-2786.

4.5 Goel P., 1978, "RAPS test pattern generator", IBM Technical Disclosure Bulletin, Vol. 21, No. 7, pp 2787-2791.

4.6 Ozguner F., Donath W.E., and Cha C.W., 1979, "On fault simulation techniques", Journal Design Automation and Fault-Tolerant Computing, Vol. 3, No. 2, pp 83-92.

4.7 Goel P., and Rosales B.C., 1979, "Test generation and dynamic compaction of tests", Proc. IEEE Test Conf., pp 189-192.

4.8 Bottorff P.S., 1980, "Test generation", Proc. NATO Advanced Study Institute on Computer Design Aids for VLSI circuit (Italy), Section 5.

4.9 Goel P., and Rosales B.C., 1980, "Dynamic test compaction with fault selection using sensitizable path tracing", IBM Technical Disclosure Bulletin, Vol. 23, No. 5, pp 1954-1958.

4.10 Goel P., et al., 1980, "LSSD fault simulation using conjunctive combinational and sequential methods", Proc. IEEE Test Conf., Paper 14.4, pp 371-376.

4.11 Goel P., 1981, "An implicit enumeration algorithm to generate tests for combinational logic circuits", IEEE Trans. Computers, Vol. C-30, No. 3, pp 215-222.

4.12 Goel P., and Rosales B.C., 1981, "PODEM-X: an automatic test generation system for VLSI logic structures", Proc. 18th IEEE Design Automation Conf., Paper 13.3, pp 260-268.

5

Practical guidelines for designing testable circuits

In this chapter, we shall present a number of practical guidelines for improving the testability of a digital chip or PCB. Many of these techniques are simply engineering commonsense and have been hinted at in earlier chapters. It is useful to collect them together, however, into a single checklist. The guidelines are classified into two types; those that aid test-pattern generation and those that aid test application and fault finding. (There does not seem to be very much that can be done to assist test-pattern evaluation, except the obvious one of providing suitable simulation support tools).

5.1   AIDS TO TEST-PATTERN GENERATION

GUIDELINE 1

Maximise the controllability and observability
features of the design

We now know that the ability to generate and apply tests is very dependent on the ease or otherwise of being able to control and to observe the values of the internal nodes of the circuit. In practice, we are not able to do this to every node but there are various techniques for improving both the controllability and observability characteristics of a design. These techniques are the subject of this guideline and we will consider, first, what we are trying to achieve and, second, how to implement the requirement.

In terms of what we are trying to achieve, consider controllability. A node is controllable if the tester is able easily to control the value of the node to any desired level (0, 1 or high impedance). Similarly, a node is observable if the tester is able easily to observe the correctness of the value of node.

If nodes are easily controlled and observed then it follows that internal logic devices are also easily controlled and observed. Examples of key control points in a logic circuit are:

(a) CLOCK and PRESET/CLEAR inputs to stored-state devices such as flip-flops, counters and shift registers;
(b) DATA SELECT inputs to multiplexer and demultiplexer devices;
(c) TRISTATE CONTROL lines on devices with tristate outputs;
(d) ENABLE/HOLD inputs to microprocessors;
(e) ENABLE and READ/WRITE inputs to memory devices;
(f) CONTROL, ADDRESS and DATA BUS inputs on any bus-structured design.

Examples of key observation points, more commonly called test points, in a logic circuit are:

(a) any buried, i.e. not directly accessible, control lines such as those listed above;
(b) outputs from stored-state devices such as flip-flops, counters and shift registers;
(c) outputs of 'data-funnelling' devices such as parity generators, priority encoders and multiplexers;
(d) any logically redundant nodes (discussed further in Guideline 2);
(e) the trunk section of nodes of high fanout;
(f) global feedback paths (discussed further in Guideline 7).

Identification of key control and observation points can be achieved either by checklists as above or by a CAMELOT analysis (as described in Chapter 2) or by a mixture of both approaches.

Once the key points are known, the next step is to modify the basic design to improve the control or observation features. General methods for improving access are shown in Fig. 5.1 and include the use of:

(a) unused chip pins or board edge-connector positions in conjunction with spare logic gates;
(b) dual-inline package sockets with a removeable straight-line connector plug;
(c) board-mounted stake pins (terminal posts);
(d) isolation by tristate drivers with additional control and observation by stake pins;
(e) single-pin or multi-pin clips direct onto the leads of integrated-circuit devices;

In connection with point (a) above, consideration can also be given to the use of a second edge-connector, positioned along another edge of the board and providing additional tester-only access to or from the board. A disadvantage with this approach is that it increases the interface requirements between the board and the tester (contrary to Guideline 19).

Note that access points, such as stake pins and clips, can be used not only as observation points but also as limited control points by overdriving onto the node. A test point on a node between two TTL devices can be driven low (but not high) to allow partial control of the driven device. The technique is shown in Fig. 5.2 and is a useful means of controlling the flow of data around a feedback loop.

If primary access to the design is limited, multiplexers and demultiplexers can be used to improve controllability and observability characteristics, as shown in Fig. 5.3. Note that the major penalties are additional devices and additional propagation delays.

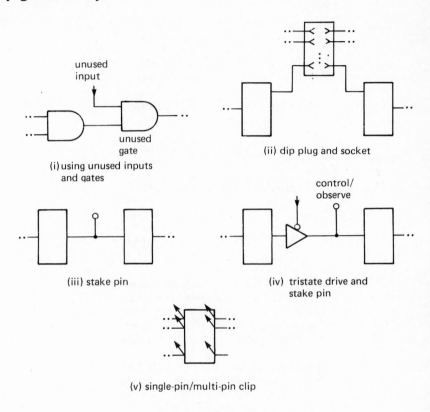

(i) using unused inputs and gates

(ii) dip plug and socket

(iii) stake pin

(iv) tristate drive and stake pin

(v) single-pin/multi-pin clip

Fig. 5.1 Methods for improving access

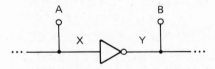

Node X is driven low through test point A. The value of node Y can now be set to either a 1 (natural state) or a 0 (driven low through test point B). The value of node X can be observed through test point A with or without node Y held low through test point B.

Fig. 5.2 Extra control by overdriving through test points.

118

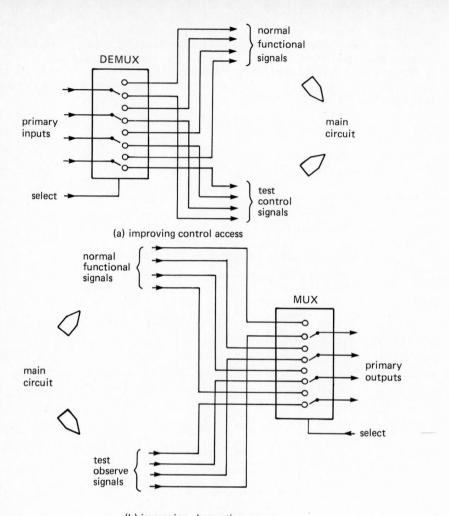

Fig. 5.3 Using multiplexers and demultiplexers.

If access is extremely limited, a parity generator can be used instead of the multiplexer shown in 5.3(b), to channel information off the circuit. A parity generator has the unique property that any single input change will cause the output to change value. If the test-observation points are chosen with some care, the output of the parity generator can be used not only to detect the presence of the fault but also to help locate the failing node.

Figure 5.4 shows other ways of using multiplexers, this time to increase the CY or OY values of low-valued internal nodes.

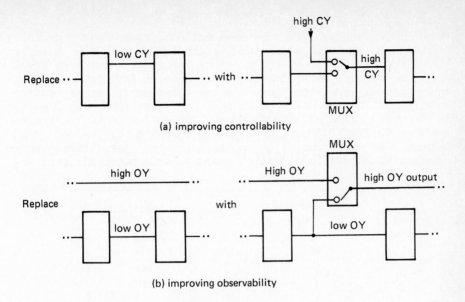

(a) improving controllability

(b) improving observability

Fig. 5.4 Improving low nodal CY and OY.

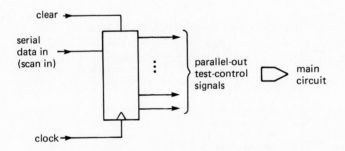

(a) improving control access (serial-in, parallel-out)

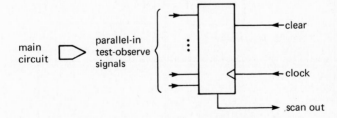

(b) improving observation access (parallel-in, serial-out)

Fig. 5.5 Using shift registers.

Shift–registers can play a similar role in improving access, as shown in Fig. 5.5. Figure 5.5(a) shows a serial–in, parallel–out shift register used to set up a particular test state. (The register is sometimes called a test–state register.) Figure 5.5(b) shows a parallel–in, serial–out register used to collect information about the state of the circuit prior to strobe out. The reader will notice some familiarity between these configurations and the more formal scan–design configurations described in Chapter 3. Indeed, the configurations in Fig. 5.5 can be categorised as scan designs. Figure 5.6 shows how the two functions in Fig. 5.5 can be merged to form a configuration known as Scan Set and based on a parallel–in, parallel–out shift register, with additional serial scan in, scan out facilities.

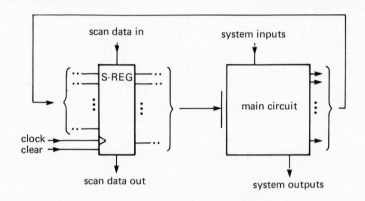

Fig. 5.6 Scan–set configuration.

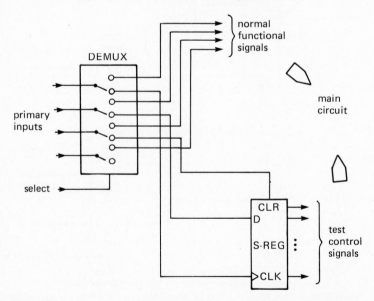

Fig. 5.7 Using a demultiplexer and shift register.

The major differences between Scan Set and the more formal scan designs of Chapter 3 are that Scan Set allows the main circuit to be of any type, not restricted to combinational, and that the stored-state devices are not restricted to specific latch or flip-flop designs. Generating a suitable set of tests for the main circuit may still be a complex task.

Figure 5.7 shows how demultiplexers and shift registers can be used together in situations in which access is very limited but for which there is a need for substantial test control signals. Undoubtedly other configurations exist. These are left to the ingenuity of the reader!

Finally, as this book was going to press, AMD Inc. announced a new Serial Shadow Register device, the Am29818. This device contains both an operational 8-bit shift register and an 8-bit shadow register. The shadow register can either be pre-loaded independently or be loaded with the current contents of the operational register. Similarly, the operational register can either be loaded independently or be loaded with the contents of the shadow register. Effectively therefore, the shadow register is designed as a scan register and is used to pre-set values into the operational register and to accept and make visible the contents of the operational register.

Because of the importance of this new type of device, and others to follow, the Appendix of the book contains a reprint of part of the data sheet. It is devices like this which, at last, put the full power of scan design techniques at the disposal of the board and system designer.

GUIDELINE 2

Avoid logical redundancy

A circuit node is logically redundant if all the output values of the circuit are independent of the binary value on the node for all input combinations or state sequences.

Logical redundancy often exists in circuits either intentionally, for example to mask a static-hazard condition, or unintentionally. The problem with a logically-redundant node is that, by definition, it is not possible to make a primary output value dependent on the value of the redundant node.

This means that certain fault conditions on the node cannot be detected, thus creating two problems. The first is that the fault condition may reintroduce the hazard condition it was designed to eliminate; the second that the fault on the redundant node may mask the subsequent detection of a second fault on a non-redundant node. Figure 5.8 illustrates the first possibility.

In this example, gate G3 is included to eliminate the static hazard possibility when switching from the (X1X2) term to the $(\overline{X1}X3)$ term, i.e., G3 provides the 'bridging' term. From a

Boolean point of view, however, the output of G3 is logically
redundant. Unfortunately it is not possible to propagate the
effect of a s–a–1 fault on the output of G3 through gate G4. The
conditions to set G3 output to 0 are inconsistent with those
required to set G1 and G2 outputs to 1. Hence a sensitive path
through G4 cannot be established. What this means is that the
circuit will continue to operate correctly but with the
possibility of a static–1 hazard condition (negative glitch) on G4
output as X1 changes from 1 to 0 with X2 and X3 held at 1. The
width of this pulse will be determined by the difference in the
signal propagation paths and may or may not be wide enough to
preset the flip–flop. It is important therefore that direct
observation of the output of G3 is made possible so that the s–a–1
fault can be tested.

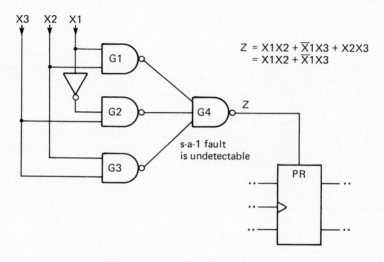

Fig. 5.8   Undetectable fault reintroducing a hazard condition.

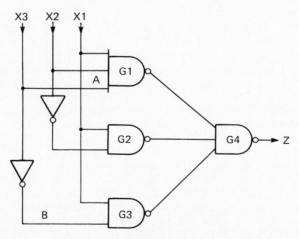

Fig. 5.9   Fault masking.

The second possibility is that a non-detectable fault on a redundant node can mask the detection of a normally-detectable fault. An example of this is shown in Fig. 5.9 in which the fault A s-a-1 is undetectable at Z because of the incompatible requirements for the values of X1, X2 and X3.

The fault B s-a-0 is detected at Z by the test input X1 = 1, X2 = 1, X3 = 0. (This is the only test for this fault.) The presence of A s-a-1, however, prevents detection of B s-a-1.

GUIDELINE 3

Keep analogue and digital circuits physically apart

The tester requirements and test strategies for analogue circuits are substantially different from those for digital circuits and the two types of circuit should be kept physically apart, even if they exist on the same PCB or even the same device, e.g. codec and digital filter chips. This is because the sharp edges of digital signals can cause crosstalk problems if they are close to analogue lines. If digital signals have to be routed in close proximity to analogue lines, then the digital lines should be properly balanced and shielded transmission lines.

It is also convenient if analogue signals that are inputs to analogue-to-digital converters are brought out for observation prior to their conversion. Similarly, digital inputs to digital-to-analogue converters should be observable as digital signals. In this way, the analogue and digital sections can be tested separately and with different test equipment if necessary.

GUIDELINE 4

Partition large circuits into smaller sub-circuits to reduce test-generation effort

---

For an SSI/MSI-based PCB containing n devices, a rule of thumb for test-pattern generation is that the amount of effort to produce and fault-simulate the tests is proportional to somewhere between n^2 and n^3. If the circuit can be partitioned into two sub-circuits (for test purposes at least), then the amount of effort is reduced correspondingly. For example, for n = 100 and assuming the cube relationship, the reduction in effort for two sub-circuits each containing 50 devices is:

$$(50^3 + 50^3) : (100^3)$$

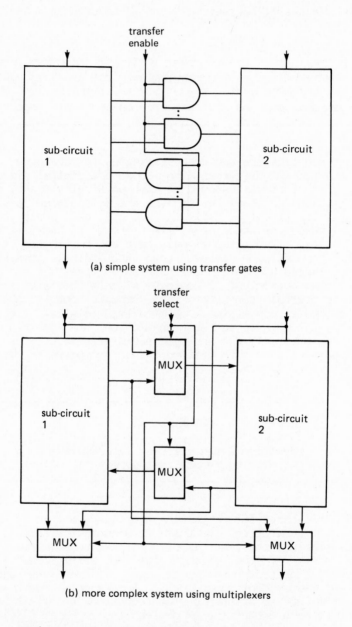

(a) simple system using transfer gates

(b) more complex system using multiplexers

Fig. 5.10 Partitioning into sub–circuits.

i.e. a reduction to approximately one-seventh of the original effort. The advantage of this reduction becomes even more noticeable if the design is changed some time after the original test program has been completed.

Logical partitioning of the circuit should be based on recognisable sub-functions and can be achieved physically by incorporating facilities to isolate and control clock lines, reset lines and even power-supply lines. Alternatively, one section of a circuit can be separated logically from another by means of tristate buffers, control gates or multiplexers as shown in Fig. 5.10.

To illustrate how partitioning helps the test-generation exercise, consider the circuit diagram for the SN7480 gated full adder, used earlier in Chapter 4 and shown again in Fig. 5.11.

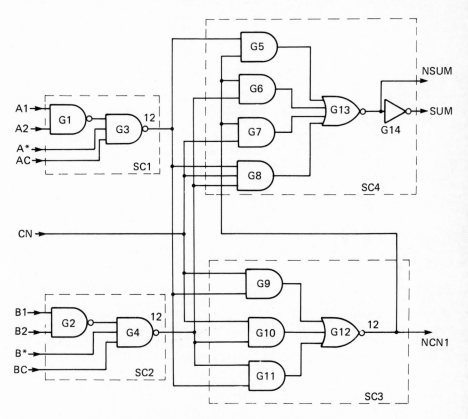

Fig. 5.11 SN7480 gated full adder.

The circuit partitions naturally into four sub-circuits, SC1-SC4, as shown. Consider now how to test the circuit in terms of the four sub-circuits.

(a) To test SC1. Generate a set of tests for SC1 assuming G3.12 to be the output. Make the value of G3.12 visible at NCN1 by setting CN = 0 and G4.12 = 1 (i.e. B* = 0). A suitable set of tests for SC1 covering all single s–a–1, s–a–0 faults is given by:

A1	A2	A*	AC	CN	B*	NCN1
0	1	1	1	0	0	1
1	0	1	1	0	0	1
1	1	1	1	0	0	0
0	0	0	1	0	0	0
0	0	1	0	0	0	0
0	0	1	1	0	0	1

(b) To test SC2. Similar to SC1. Tests are:

B1	B2	B*	BC	CN	A*	NCN1
0	1	1	1	0	0	1
1	0	1	1	0	0	1
1	1	1	1	0	0	0
0	0	0	0	0	0	0
0	0	1	0	0	0	0
0	0	1	1	0	0	1

(c) To test SC3. Generate a set of tests assuming G3.12, CN and G4.12 to be the inputs and with visibility at NCN1. Note that the primary input, CN, and the pseudo–inputs, G3.12 and G4.12, can each be set to 0 or 1 individually and independently. A suitable set of tests is given by:

A1	A2	A*	AC	B1	B2	B*	BC	G3.12	G4.12	CN	NCN1
0	0	1	1	0	0	1	1	0	0	1	1
0	0	1	1	–	–	0	–	0	1	0	1
0	0	1	1	–	–	0	–	0	1	1	0
–	–	0	–	0	0	1	1	1	0	0	1
–	–	0	–	0	0	1	1	1	0	1	1
–	–	0	–	–	–	0	–	1	1	0	0

(d) To test SC4. This one is more interesting insofar as one of the pseudo–inputs, G12.12, is itself a function of the other three inputs (G3.12, CN, G4.12). This means that certain combinations of all four inputs are not achievable. The reader should verify that the following set of tests is achievable and covers all single s–a–1, s–a–0 faults in SC4:

A1	A2	A*	AC	B1	B2	B*	BC	G3.12	G4.12	CN	G12.12	SUM
0	0	1	1	0	0	1	1	0	0	1	1	1
0	0	1	1	–	–	0	–	0	1	0	1	1
0	0	1	1	–	–	0	–	0	1	1	0	0
–	–	0	–	0	0	1	1	1	0	0	1	1
0	0	1	1	0	0	1	1	0	0	0	1	0
–	–	0	–	–	–	0	–	1	1	0	1	1

Even without any test compaction, the circuit is fully tested for single s–a–1, s–a–0 faults by 24 tests. This figure compares very favourably with the 512 $(= 2^9)$ tests required to test the circuit exhaustively, and even with the 48 tests required to test each sub-circuit exhaustively. $(48 = 2^4$ for SC1 + 2^4 for SC2 + 2^3 for SC3 + 2^3 achievable for SC4).

The reader is invited to carry out a similar exercise for the SN74181 Arithmetic Logic Unit device. (See reference 5.19 for a partial solution for this device.)

GUIDELINE 5

Avoid asynchronous logic

Asynchronous logic employs stored-state devices (in the form of latches) and global feedback but the state-transitions are governed solely by the sequence of changes on the primary inputs. There is no system clock to determine when the next state of the circuit will be established.

The advantage of asynchronous circuits is speed of operation. The speed at which state-transitions occur is limited only by the propagation delay of the gates and interconnects. In this respect, the design of asynchronous logic is more difficult than synchronous (clocked) logic and must be carried out with due regard to the possibility of races. Design techniques to eliminate race conditions do exist – the problem is that the presence of faults may reintroduce the race. Providing test patterns for such circuits can prove very difficult, particularly if the outcome of the race is not deterministic, i.e., in the critical race situation. Not only is the circuit difficult to test but the possibility of non-deterministic behaviour can cause problems during fault simulation. As a result, synchronous logic is preferable to asynchronous logic even to the point of using synchronous counters rather than asynchronous ripple counters or even asynchronously-coupled synchronous counters.

GUIDELINE 6

Ensure that stored-state devices are easily initialised

Initialisation is a necessary precursor to any practical test program and simulation run. Ideally it should be possible to set every stored-state device in the circuit into a known start state.

128

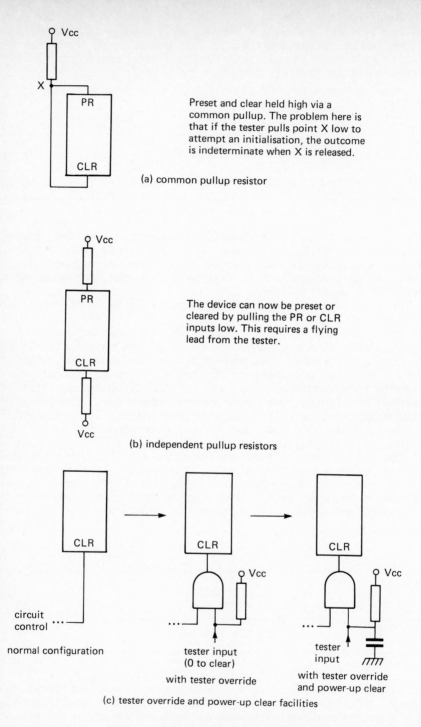

Preset and clear held high via a
common pullup. The problem here is
that if the tester pulls point X low to
attempt an initialisation, the outcome
is indeterminate when X is released.

(a) common pullup resistor

The device can now be preset or
cleared by pulling the PR or CLR
inputs low. This requires a flying
lead from the tester.

(b) independent pullup resistors

circuit
control

normal configuration

tester input
(0 to clear)

with tester override

tester
input

with tester override
and power-up clear

(c) tester override and power-up clear facilities

Fig. 5.12 Initialisation of stored–state devices

What sometimes happens, at PCB level, is that individual preset or clear lines are tied to Vcc through a pullup resistor. Figure 5.12 illustrates some of the problems of, and solutions to, this particular requirement. Note that tester override is necessary not only as a means of providing direct control on the CLR input but also to allow a test of the power-up facility itself. It would also be extremely inconvenient if the only way of initialising the flip-flop is by switching the power off and then back on to activate the action of the RC circuit!

Other problems of initialisation can occur if the state of one device is dependent on the state of another, as in an asynchronous ripple counter, for example. If independent control of each flip-flop in the counter is not provided, initialisation can only be achieved by clocking the counter until a particular state is established and identified by the tester. Even when direct access to a master RESET is provided, the designer can still fall into an initialisation trap. Figure 5.13 illustrates one such situation. The reader is invited to derive an initialisation sequence for what is really a very simple counter circuit.

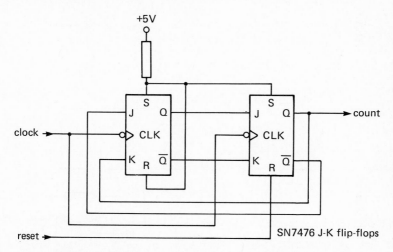

Problem: Initialise the counter to a known start state

Fig. 5.13 Counter with initialisation problem.

GUIDELINE 7

Provide facilities to break feedback paths

The importance of this facility cannot be overstressed. Global feedback paths complicate both test generation and fault

130

diagnosis. Feedback loops can be broken and controlled by a variety of techniques, as shown in Fig. 5.14.

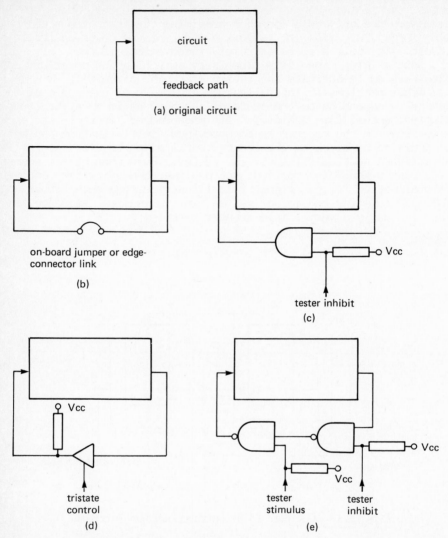

(a) original circuit

on-board jumper or edge-connector link

(b)

tester inhibit

(c)

Vcc

tristate control

(d)

tester stimulus

Vcc

tester inhibit

(e)

Fig. 5.14 Techniques for breaking and controlling feedback paths.

GUIDELINE 8

Avoid the use of monostables

There are various testing problems associated with the use of monostables in logic circuits.

The first is that direct observability of the monostable output is necessary if the period of the monostable is to be tested. If it is not possible to make the output directly observable then the test programmer will have to resort to the use of a clip.

The second problem is that even if the output is observable, the period may be too fast for the tester. This problem can be solved either by the use of a follow-on latch to catch the pulse or by a capacitor clip over the monostable capacitor to lengthen the period. These techniques are illustrated in Fig. 5.15.

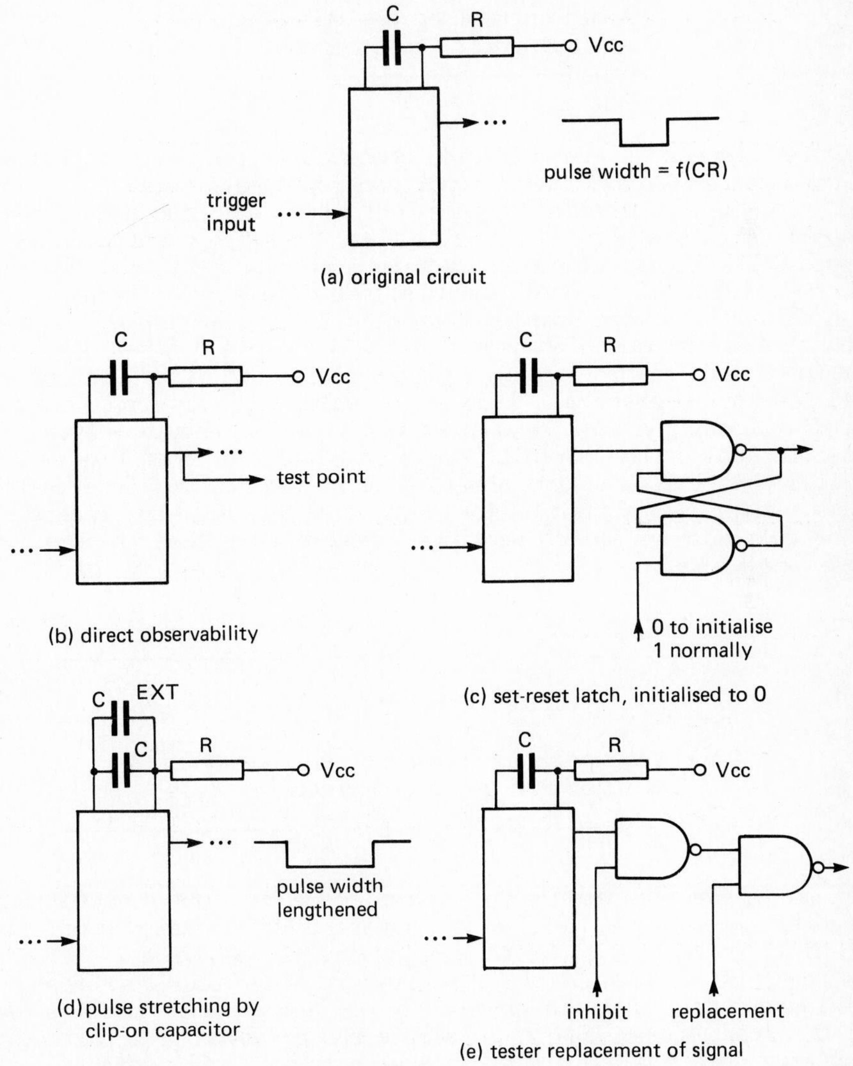

(a) original circuit

(b) direct observability

(c) set-reset latch, initialised to 0

(d) pulse stretching by clip-on capacitor

(e) tester replacement of signal

Fig. 5.15 Techniques for handling monostables.

Conversely, a monostable may have a long period (of the order of ms or even seconds). The problem now is not how to measure the width of the pulse but how to shorten the pulse so that the period does not slow the testing speed. This can be achieved by a resistive clip over the monostable resistor.

Finally, if the use of a monostable cannot be avoided, then direct access to the reset line must be provided.

GUIDELINE 9

Avoid customised designs using 'singular' components

Singular components are circuit components that are specially selected or adjusted to suit conditions on a particular PCB. They are commonly termed 'select-on-test' or 'adjust-on-test' components. Examples are multi-turn locking potentiometers or tapped delay lines. The problem with such components is that it is often difficult to establish a standard test or, alternatively, that the test is more complex than need be. For example, a test program can include an automatic test to measure a fixed delay but will have to resort to a program loop and the use of an oscilloscope to observe a delay whose value can vary from board to board. If designs must be customised, then it should always be possible to set the circuit into a standard test condition. This implies that customisation should be achieved not by irreversible means, such as cutting links or tracks, but by reversible techniques such as switch settings, removable socket blocks, or backplane links.

GUIDELINE 10

Avoid test-program dependency on the contents of ROM and PAL devices

---

ROM and PAL devices are used as replacements for combinational circuits implemented with basic gates. They are also used, with feedback, to realise a particular sequential behaviour. From a testing point of view, however, there are other considerations.

The first is that the device may not contain the precise set of 0s and 1s necessary to enable the creation of a particular sensitive path somewhere else in the circuit. If this is the case, then the test programmer may be forced to set the device output lines to their tristate level (by controlling the ENABLE pin) and then clip over the top of the device to provide tester access to the device output pins, as shown for a ROM in Fig. 5.16.

In Fig. 5.16(a), the ENABLE line has been tied directly to 0v, i.e., the ROM is permanently enabled. This means that the tester is unable to control the line. Figures 5.16(b) and 5.16(c) show two ways of allowing the tester access to the line. In Fig. 5.16(b), the input is pulled low, whereas in Fig. 5.16(c) it is pulled high. The latter scheme has more immunity to noise but requires more components.

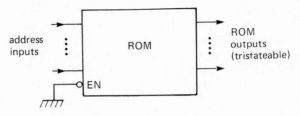

(a) on-board ROM, ENABLE grounded

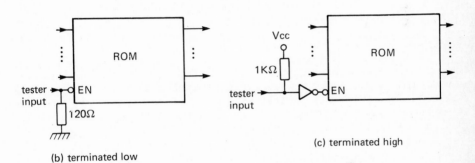

(b) terminated low

(c) terminated high

Fig. 5.16 Terminated ROM enable line.

Control of the ENABLE line and clipping of the device is important from another point of view. The contents of a ROM or PAL device are sometimes changed as the board undergoes revision. It is sensible therefore to reduce the dependency of the test program on the device contents in order to prevent major modifications to an existing test program.

GUIDELINE 11

Terminate all unused device inputs and
tristateable or open-collector outputs

From a design point of view, unused inputs to logic devices should always be terminated to Vcc or 0v through a suitable resistor.

This is done to remove the risk of noise pickup on inputs left floating.

Termination of unused inputs is also important for testing purposes in order to allow possible tester control of the behaviour of a device (as discussed in previous guidelines). It is important, however, that termination is via a resistor so that the tester can pull the terminated node high or low.

Used and unused outputs from tristate or open-collector devices should also be terminated with a pull-up resistor to prevent inconsistent logic values leading to inconsistent guided-probe nodal signatures.

If the design of the system is such that a series of bus devices are assembled on different boards and connection from one device to another is via a back-plane bus, then it is common to find unterminated tristateable outputs. (The pull-up resistors are usually mounted as a block on a bus termination board.) For this situation, a special adaptor will be required to test the board if the tester does not contain pull-up or pull-down facilities. One way to avoid this problem is to include bus pull-up resistors on each board. In normal operation, all but one set of the resistors would be made inoperative, either by physical removal or by on-board switches.

5.2 AIDS TO TEST-APPLICATION AND FAULT-FINDING

The previous eleven guidelines related mostly to problems associated with the generation of suitable test patterns. This section continues with guidelines that relate more to problems associated with applying the tests and to carrying out diagnostic procedures in the event of failure.

GUIDELINE 12

Locate equivalent faults to the same integrated circuit package

---

If fault diagnosis is primarily by correlation of the circuit response with the entries in a fault dictionary, then it is useful to group equivalent faults onto the same integrated-circuit device. In this way, there is no need to attempt a further refinement of the diagnosis by means of a guided probe. Figure 5.17(a) contains an example of this.

In this circuit, a s-a-1 fault on G1.1 is tested by the same set of tests as for the equivalent pair G2.1 s-a-0 and G2.2 s-a-1 around the inverter G2. If there is any choice therefore, the entire circuit should be implemented with gates from the same quad 2-input NAND gate package, Fig. 5.17(b), rather than with three NAND gates from one package and one inverter from another, Fig. 5.17(c).

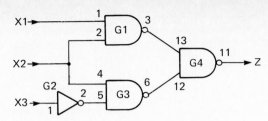

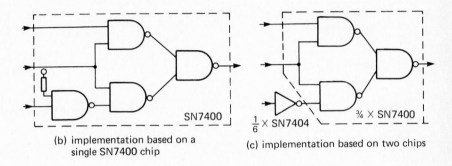

(a) G1.1 s-a-1 is equivalent to G2.1 s-a-0 and G2.2 s-a-1

(b) implementation based on a single SN7400 chip

(c) implementation based on two chips

Fig. 5.17 Effect of equivalent faults on diagnostic resolution.

In practice, there is still the need to identify the fault source down to the actual node in order to eliminate the printed-circuit track as the cause of the fault. This requirement tends to invalidate the recommendation but, under certain circumstances, grouping equivalent faults can be useful, i.e. where it is known that a particular device is susceptible to failure. This is particularly true for a set of gates whose outputs are wired together, either as a wired-OR or wired-AND.

GUIDELINE 13

Allow the tester direct control of clock circuitry

For circuits containing free-running on-board oscillators, it is usually necessary to replace the internal clock with one generated externally by the tester. In this way, the speed of operation of the circuit can be reduced, down to single-step operation if necessary. This has value not only for test application but also during the debug phase of test generation. Techniques for replacing internal clocks are shown in Fig. 5.18.

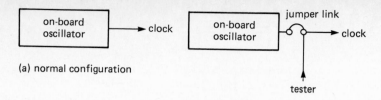

(a) normal configuration

(b) jumper or edge-connector link

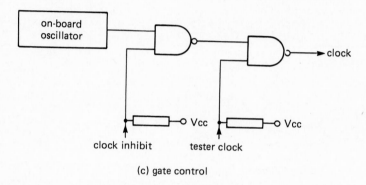

(c) gate control

Fig. 5.18 Techniques for replacing internal clocks.

GUIDELINE 14

Avoid diagnostic ambiguity groups such as wired–OR
wired–AND junctions and high fanout lines

Wired–OR, wired–AND junctions and high fanout lines present
problems of ambiguity of fault diagnosis. If possible, they
should be avoided. (See Fig. 5.19.)

It is also useful if gates whose outputs are wired together
are confined to a small number of packages. In this way, if
several gates are damaged because of a fault on the wired output
connection, then only a few devices (preferably one) need be
changed.

GUIDELINE 15

Break up long counter chains

Figure 5.20(a) shows a 16–bit counter constructed from two 8–bit counters.

In order to test the counter fully, $2^{16} + 1 = 65537$ clock pulses should be applied. At a test rate of 50 kHz, this represents a test time of approximately 1.3 seconds. Now consider the modified version shown in Fig. 5.20(b). In this circuit, each 8–bit counter can be tested separately and the total test time is reduced to $2 \times (2^8 + 1) \times 0.02$ ms, i.e. approximately 10 ms. The saving in time is important and is obviously valuable in its own right, but it is also useful if there are subsequent requirements to set the counter to a particular count for tests associated with other devices on the board. (In this respect, independent access to the pre–load facilities on each counter device is very useful.)

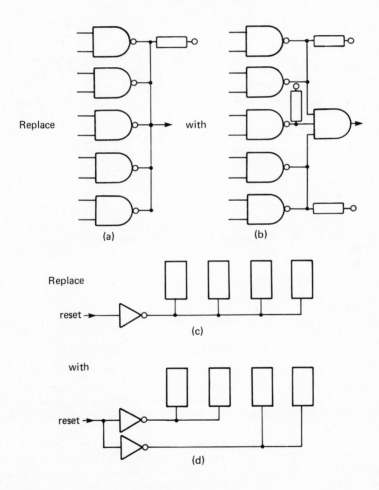

Fig. 5.19 Reducing diagnostic resolution ambiguity.

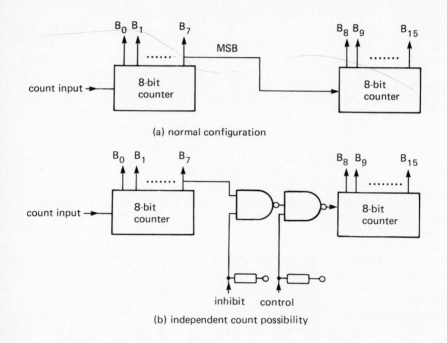

(a) normal configuration

(b) independent count possibility

Fig. 5.20 Breaking up long counter sequences.

GUIDELINE 16

Buffer edge-sensitive input signals

Many control signals to stored-state devices, such as clock signals, specify minimum rise or fall times. If these signals come direct from an edge-connector position, then it may be necessary to incorporate a buffer device (line driver, Schmitt trigger, or two inverters) in the interface between the board and the tester to speed up the rise and fall times of tester-generated signals. The need to do this is dependent on the slew rate of the driver amplifier in the tester. Effectively what this guideline says is that devices that require fast edges should be kept away from the primary inputs and primary outputs.

GUIDELINE 17

Take care with board layout and construction to
reduce operator fatigue and provide physical access

To help reduce operator fatigue when fault-finding with a
hand-held guided probe over a range of different board types, it
is helpful if device placement is regular (in columns and rows)
and device orientation is standard, i.e. pin 1 is always
top-left-hand corner, say. Also, mis-probes can occur as the
operator moves from a 14-pin device to a 16-pin device, although
it is difficult to suggest a solution to this problem.
Possibilities are to mark all 16-pin devices with a colour marker
or to contain such devices into one or two columns on the board.
    Edge-connector types should be standardised (e.g. same pitch)
and power and ground supplies should always be at the same
positions. Bus inputs should be physically adjacent and lined up
to suit the tester driver/sensor pin groupings. (Some testers are
only able to change groups of 16 pins, say, at a time in
parallel.) There should also be clear identification of the board
number and modification level, of the grid reference on the board,
and of any discrete components that are identified by a special
identifier (R1, R2, etc.) rather than by their grid reference.
    For boards that are to be diagnosed by guided probe, it is
helpful if all devices and printed-circuit tracking are visible
from a single side of the board (although this is not possible
with multi-layer boards). The visibility of devices is necessary
for conventional probing and the visibility of track is necessary
for current-sensing probes. In this context, it is not possible
to follow copper track that is routed beneath an integrated
circuit device. Similarly, plated-through holes (vias) can create
tracing problems. For any form of contact probing, non-conducting
protective coatings are to be avoided.
    On-board test points that are not brought out to
edge-connector positions should be gathered close together to
simplify the interface requirement of flying leads. This can be
done by using a dummy integrated-circuit socket or by centralising
all points onto a series of regular-pitch stake pins or wire-wrap
pins. Access is then via a single harness terminated with a
suitable plug or socket.
    There may be a requirement to gain access to a device by means
of a single-pin or multi-pin clip. If this is likely, then care
should be taken to ensure that there is enough room to accommodate
the clip. Logic devices are not usually located too closely
together, but occasionally discrete components such as pull-up
resistors or decoupling capacitors are located close enough to a
device to prevent the use of the clip.
    Finally, it is often thought advisable to mount complex

140

devices in sockets so that they can be removed during the
application of the tests and replaced if found to be faulty.
These advantages are offset by the increased risk of bent leads
during reinsertion, wrong orientation, wrong chips replaced, and
longer test set-up times. If these risks outweigh the advantages
of socketted devices, it is better to solder the device onto the
board but to provide a means of electrical isolation. In any
case, the go/no go part of the test program should not require
removal of the device, even if it is socketted.

GUIDELINE 18

Provide a short–circuit link to check
board alignment on the tester

It is good practice for the operator to check the alignment and
orientation of the board before running the program, but it is
helpful if the program itself can carry out a check before
applying power. One way to do this is to make use of a pair of
unused edge-connector positions and to short-circuit between these
positions on the board. The test program can then test that the
short-circuit exists before continuing with the program. (See
Fig. 5.21.)

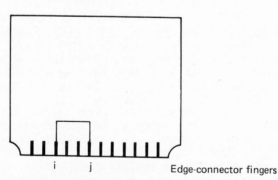

i   j   Edge-connector fingers

Edge-connector fingers, i and j, have been shorted together on the board to allow the test
programmer to include a test for correct alignment of the board prior to the application
of power. In this way, the possibility of damage (to board or tester) due to mis-alignment
is considerably reduced.

Fig. 5.21 Automatic testing of board alignment.

GUIDELINE 19

Keep the board-tester interface simple

The standard interface between a board and a tester consists of the primary input, primary output stimulus/response interface, plus power supply distribution. In addition, there may be other specialised interface requirements such as 'flying lead' probes (single or multi-pin), buffer devices, clipped components, connectors on more than one side of the board, multiple connector types, and on-board connectors. Although each of these facilities may be provided to ease a particular testing problem, they represent an additional complication from the test-engineer's point of view and, in a distributed testing environment, from a cost and logistics point of view. The leads can also represent a considerable capacitative load on the board. The general engineering maxim of 'keep it simple' applies just as much here as elsewhere.

GUIDELINE 20

Avoid mixed logic families on the same board

Different logic families require different threshold reference values on driver/sensor pins and, usually, different power-supply levels. The use of two or more logic families on the same board can therefore cause various complications to the interface. One solution, if mixed logic is unavoidable, is to ensure that all primary inputs and primary outputs are at least TTL compatible.

GUIDELINE 21

Limit device fanout to one less than maximum

If a device output is being used to drive its maximum load, then the addition of a guided probe on the node may just be enough to overload the device and cause degradation of the signal. To avoid this problem, high-fanout devices should be limited to drive one less than their design maximum.

142

GUIDELINE 22

Provide on-board 'stay-alive' circuitry

Provide 'stay-alive' facilities, such as MOS RAM refresh cycles, on the same board as the devices that require it. This removes the need for the tester to provide the facility, the problems being:

(a) in terms of hardware, the tester may not be fast enough both to provide the refresh cycle and carry out useful tests, and
(b) in terms of test program software, the need to incorporate regular calls to refresh subroutine procedures.

GUIDELINE 23

Know the limitations of the target ATE

Ultimately, the chip or board has to interface electrically with the tester. It is therefore important to know and understand the limitations of the tester. Important features are as follows.

(a) Maximum speed at which a driver/sensor pin can be changed from driver mode to sensor mode. This is important for driving and sensing bidirectional lines.
(b) Whether dynamic events, such as fast pulses, can be captured and measured. This is termed 'pulse catching'.
(c) Driver pins can usually be changed either one pin at a time (skew mode) or collectively (broadside mode). The normal mode is skew but, occasionally, broadside mode is required – for example, when driving information onto a set of bus lines. Even in broadside however, there will be a certain amount of skew between the pin changes. This skew may be significant for fast bus changes.
(d) Driver pin changes from 0-to-1 and from 1-to-0 will have measurable rise and fall times usually fixed by the slew rate of the driver operational amplifier. These times may be different for individual driver pins, and in any case, fix the maximum speed at which value changes can be made.
(e) Sensor pins operate with single or dual reference threshold values to identify logic 0 and logic 1 values in the circuit-under-test. These threshold values should obviously

be programmable and, in some cases, it will be necessary to re-define the values during the course of the test program.

(f) If the circuit-under-test contains a free-running oscillator, then it is useful if the operation of the tester can be synchronised to the frequency of the oscillator. A 'sync' facility will enable this to happen.

(g) If there is a requirement for some sort of a repetitive background sequence, such as a MOS RAM refresh cycle, check that the tester can provide the sequence whilst allowing other useful tests to be carried out at the same time.

(h) The STROBE pulse is an important feature of the tester. It determines the point at which value changes are driven onto the circuit through the driver circuitry and the point at which values sensed off the board are latched by the sensor circuitry. It is necessary to be able to change the width of the pulse (to match the propagation delay characteristics of the circuit-under-test) and the frequency (to suit some requirement for timing measurement). Both the width and the frequency of the STROBE pulse will have maximum and minimum values.

(i) Driver/sensor pins will have other electrical limitations such as: maximum current source and sink capabilities; maximum positive and negative voltage swings; voltage margins for both driving and sensing logic 0s and logic 1s; load impedance characteristics (check the capacitive loading of interface leads as well).

(j) Check the availability of resistive pull-up, pull-down facilities of driver/sensor pins. These facilities are required for testing the high-impedance status of tristate and open-collector lines.

GUIDELINE 24

Pay special attention to boards that are
to be tested on an in-circuit tester

In-circuit testing is currently a useful and popular method for identifying defects on loaded printed-circuit boards caused by the manufacturing assembly processes. The purpose of the in-circuit tester is to test each device on the board, analogue and digital, in isolation from all other devices on the board. Electrical access to the devices is by way of a bed-of-nails fixture which provides a direct contact between the tester's driver/sensor pins and the non-component side of the board. The bed-of-nails pins make contact either directly with solder joints or with copper contact pads connected to the device pins. Passive analogue components are isolated electrically by a technique called 'guarding' (see references 5.21 or 5.27). Digital devices are isolated by back-driving into the devices that connect to the device-under-test in a manner similar to the technique of overdriving through test points described earlier in the chapter (Fig. 5.2 q.v.). The test methodology is to test first for opens and shorts (unpowered continuity test) and then to power-up the board and test each device individually against a set of predefined standard tests. Any failure at this stage is judged to be caused by the device-under-test or its immediate electrical neighbours.

The sole purpose of the in-circuit tester is to screen the whole board for manufacturing defects. The tester is not capable of testing the overall functionality of the board i.e. how the devices on the board function collectively. This is the job of the subsequent functional edge-connector tester or system substitution rig tester.

Because of the importance of the in-circuit tester, it is worth considering any special design or manufacturing features to aid testability. Many of the previous guidelines still apply of course (as will be indicated), but there are several points which relate only to boards that are to be tested on an in-circuit tester. These points are listed below. Mostly they relate to maximising access to on-board devices by paying attention to the physical layout and construction of the board; others are more to do with improving the electrical test environment.

(a) Make sure all the components are mounted on the same side of the board. Vacuum-operated bed-of-nails probing becomes more difficult, if not impossible, if there are components on the probing side.

(b) Tooling holes must be positioned to allow accurate alignment of the bed-of-nails fixture. Many of the problems of actually using in-circuit testers stem from mis-aligned fixtures causing unreliable electrical contact.

(c) In the case of a contact through a copper pad, the pad should be located very close to the device pin to which it is connected in order to reduce path lengths and loading effects. A fast transition on a back-driven node can reach the input of a device-under-test and cause an unexpected response resulting in test failure. This likelihood is particularly true for high-speed low noise immunity logic such as ECL in which every signal-transmission path must be treated as a transmission line.

(d) The particular way a device is used in a circuit should not preclude the use of device features that might be helpful in a test. Often, the full functionality of a device is not exploited in the circuit - inputs are tied off, gates are left unused. The means by which inputs are tied off should not prevent independent control if required (as already discussed in Guideline 11). Also, the pins of unused devices or gates should be connected through to the non-component side of the board and made available for bed-of-nail contact.

(There is an interesting philosophical question here.

Should the in-circuit tester only test the device as it has been configured for use in the circuit, or should the tester attempt a full device test irrespective of how the device is to be used? A 4-bit binary counter (SN74161 say) may be configured to work only from its all-0 initalised state i.e. the ability to preset to any initial count is not used – the four data lines are tied off. If the counter passes all tests except the pre-load test, should the counter be considered to have passed or failed? The answer is complex but is related to the relationship between the perceived failure and the real failure mechanism, and to whether the objective of testing is to discover all failure mechanisms or only those failures that can affect correct operation.)

(e) Take care not to fragment the ground structure too much. Under in-circuit test conditions, device outputs can carry abnormally high currents which can inject noise spikes onto the ground plane. These spikes can cause unwanted transitions elsewhere which may create what are thought to be failure conditions. This is particularly true for low noise immunity logic such as ECL.

(f) Similarly, certain buffer devices such as the SN74S37 or SN74S240 are capable of sourcing or sinking very high currents. If these devices are connected to the device-under-test, then there is a risk of electrical transients which may interfere with the test conditions and hence the outcome. It is advisable to place these devices into their 'least agressive' electrical state (output high for the SN74S37, output high impedance for the SN74S240). The circuit design should provide sufficient access to achieve this state.

Finally, many of the previous guidelines have particular application to in-circuit testers. Specifically, guideline 6 (on initialisation), guideline 7 (on breaking feedback loops), guideline 9 (on customised designs), guideline 10 (on ROM/PAL dependency), guideline 11 (on terminating unused pins), guideline 13 (on allowing direct control of clocks), guideline 17 (on other layout points such as orientation, labelling and socketing), and guideline 23 (on target ATE limitations).

GUIDELINE 25

Provide clear design engineering documentation

Last, but not least, the design engineer should provide clear documentation to support the design of the circuit. In particular, the documentation should include the following items.

(a) A complete functional specification for the circuit, including an explanation of signal names where these are meaningful.
(b) Timing diagrams and tolerances showing the results of control actions.
(c) Details of any design workarounds and notification of any parts of the design that may be modified in the future.
(d) Clear logic diagrams together with identification of major and minor feedback paths and an explanation of any unfamiliar (non-standard) logic symbols.
(e) Details of power-supply requirements, such as sequencing for multi-level supplies, maximum current drive or sink levels, any special voltage levels.
(f) ROM and PAL devices should be described by a Boolean functional expression. If this expression is not available, then a tabular hexadecimal listing of the truth-table should be provided.
(g) A recommended test strategy plus details of any self-test, fail-safe, fault-secure, or fault-tolerant facilities.

5.3 CONCLUDING REMARKS ON PRACTICAL GUIDELINES

In this chapter, we have listed and made comment on a series of practical guidelines for improving testability. The literature abounds with such lists and the chapter has tried to collect all the guidelines into a single concise anthology. The list will never be complete, neither is it true to say that designers will necessarily follow the advice. If design and test remain as separate activities within an organisation, then designers will always be more concerned with the traditional aspects of design – performance, functionality, speed, power consumption, reliability – than with testability. The only way a designer will acknowledge and conform to design-for-test is if he or she appreciates the needs and understands clearly the penalty for not adopting a design-for-test attitude. One way to do this is to merge the two activites into one by making the designer responsible for generating and evaluating a test program for the design. Only in this way will the penalties of not following a design-for-test approach become visible to the designer.

Untimately, there is no substitute for learning by experience. The purpose of the book has been to discuss various tools and techniques for designing testable logic circuits. The next stage is to put the ideas into practice.

BIBLIOGRAPHY

5.1 Friedman A.D., 1967, "Fault detection in redundant circuits" IEEE Trans. Electronic Computers, Vol. EC-16, pp 99-100.

5.2 Boswell F.R., 1972, "Designing testability into complex logic boards", Electronics International, Vol. 45, No. 17, Aug. 14, pp 116–119.

5.3 Schneider D., 1974, "Designing logic boards for automatic testing", Electronics International, Vol. 47, No. 15, July 25, pp 100–104.

5.4 Hayes J.P., and Friedman A.D., 1974, "Test point placement to simplify fault detection", IEEE Trans. Electronic Computers, Vol. C–23, pp 727–735.

5.5 Dandapani R., and Reddy S.M., 1974, "On the design of logic networks with redundancy and testability considerations", IEEE Trans. Electronic Computers, Vol. C–23, pp 1139–1149.

5.6 Writer P.L., 1975, "Design for testability", Proc. IEEE Automated Support Systems Conference, pp 84–87.

5.7 Bennetts R.G., and Scott R.V., 1976, "Recent developments in the theory and practice of testable logic design", IEEE Computer, Vol. 9, No. 6, June, pp 47–63.

5.8 Tose D., 1976–77, "Digital logic board design with test needs in mind", Electronic Engineering, Vol. 48, No. 586, Dec., pp 73–75 (Part 1); Vol. 49, No. 587, Jan., pp 46–48 (Part 2).

5.9 Fox J.R., 1977, "Test point condensation in the diagnosis of digital circuits" Proc. IEE, Vol. 124, No. 2, Feb., pp 89–94.

5.10 Mittelbach J., 1978, "Put testability into PC boards", Electronic Design, Vol. 26, No. 12, June 7, pp 128–131.

5.11 Foley G., 1978, "Designing microprocessor boards for testability", Proc. IEEE Semiconductor Test Conf., pp 176–179.

5.12 Davidson R.P., 1979, "Some straightforward guidelines help improve board testability", Electronic Design News, May 5, pp 127–129.

5.13 Boney J., and Rupp E., 1979, "Let your next microprocessor check itself and cut down your testing overhead", Electronics Design, Vol. 27, No. 18, Sept. 1, pp 100–105.

5.14 Williams T.W., and Parker K.P., 1979, "Testing logic circuits and designing for testability", IEEE Computer, Vol. 12, No. 10, Oct., pp 9–21.

5.15 Lippman M.D., and Donn E.S., 1979, "Design forethought promotes easier testing of microcomputer boards", Electronics International, Vol. 52, No. 2, Jan. 18, pp 113–119.

5.16 Hayes J.P., and McCluskey E.J., 1980, "Testability considerations in microprocessor-based designs", IEEE

Computer, Vol. 13, No. 3, March, pp 17–26.

5.17 Grason J., and Nagel A.W., 1981, "Digital test generation and design for testability", Journal Digital Systems, Vol. 5, No. 4, pp 319–359.

5.18 Williams T.W., and Parker K.P., 1983, "Design for Testability – a survey", IEEE Trans. on Computers, Vol.C–31, No.1, pp 2–15.

5.19 McCluskey E.J., 1981, "Design for autonomous test", IEEE Trans. Computers, Vol. C–30, No. 11, pp 866–875.

5.20 Chandramouli R., 1982, "Designing VLSI chips for testability", Electronics Test, Vol. 5, No. 11, pp 50–60.

5.21 Bennetts R.G., 1982, "Introduction to digital board testing", Crane–Russak (New York), Edward Arnold (London).

5.22 1983, "Initializing sequential circuits", Electronics Test, Vol. 6, No. 2, pp 25–26.

5.23 Willis R., 1983, "Design for testability for microprocessor boards", Electronic Engineering, Vol. 55, No. 676, April, pp 67–72.

5.24 Hewlett–Packard, "Designing digital circuits for testability", Application Note 210–4.

5.25 GenRad, "How to design logic boards for easier automatic testing and trobleshooting", Technical publication.

5.26 Computer Automation, "Design for testability", Technical publication (AN–104–378–500).

5.27 Friedman D., 1982, "Understanding and successfully implementing in–circuit testing", Fairchild Technical Paper, T1.

5.28 DeSimone S., 1982, "In–circuit testing of ECL loaded boards", Fairchild Technical Paper, T2.

5.29 Scheiber S.F., Vaidyanathan M.M, and Longendorfer B., 1982, "In–circuit testing – testability requirements", Fairchild Technical Paper, T3.

5.30 Fichtenbaum M., 1983, "Testability requirements for in–circuit testers", private communication.

6

Testable design
— the future?

Chapter 1 has outlined the major problems associated with testing
digital devices and systems and has discussed the need for an
awareness of testability at the design stage. Chapters 2 – 5
inclusive have addressed themselves to specific topics aimed at
solving some of the problems. It is appropriate at the end of the
book to ask the question – what of the future? How will the
demands on testing change and what effect will these changes have
on testable design policies?

In this chapter, we will summarise the present status and
attempt some forecasting of the future.

6.1 PRESENT STATUS AND TRENDS

We will consider the present status in terms of the three
categories listed in Chapter 1: test generation, test evaluation
and test application. In reality, the three activities are very
inter-related.

6.1.1 Test generation

For test generation, the major problem has been to develop a
practical general purpose test-generation system to support the
test requirements of logically-complex devices and boards. It is
now thought, by the practitioners at least, that such a system can
never be fully automatic unless the design is constrained in some
way so as to reduce the complexity of the task. In particular, if
automatic test generation is required only for combinational
circuits, then the problem is considered to be soluble (but see
later for further comment). This reasoning has been the driving
force behind the development of the various scan-path techniques
(Chapter 3) and associated combinational test-generation
procedures such as PODEM (Chapter 4). In addition, there are
certain practical constraints which assist in reducing the costs
of test generation and these have been discussed in Chapter 5.

For sequential circuits not constrained by scan design,
testability measures such as CAMELOT (Chapter 2) have been
developed to at least provide an early insight into potential

test-generation problem areas. At best, however, current testability measurement tools can only provide a coarse assessment of testability.

Within the custom LSI chip manufacturing industry, many companies are now actively considering the use of some form of self test such as the BILBO techniques described in Chapter 3. The objective is to reduce overall production costs by reducing the testing time. The penalty is increased silicon area (upto 2% extra for the microsequence tester built into the Motorola MC68000 microprocessor - reference 6.2).

Lower down the scale, gate-array and standard-device manufacturers are beginning to make use of scan-design principles e.g. the AMD Am29818 device described in the Appendix and the UK 5000 gate-array project described in reference 6.9. As system designers become more aware of chips and design techniques such as these, then so the techniques of testable design will continue to be used and improved.

But all this is short term - what of the future? Goel (reference 6.1) has derived the following results for projecting the costs of testing large 'coupled' combinational circuits i.e. combinational circuits which cannot be partitioned into disjoint substructures. G is the equivalent gate count of the circuit.

Result 1. The number of tests to achieve 100% single struck - fault cover is proportional to G

Result 2. The cost of generating these tests is proportional to G^2

Result 3. The computation costs associated with the use of a parallel fault-simulator to assess fault cover is proportional to G^3

Result 4. The computation costs associated with the use of a deductive fault simulator to assess fault cover is proportional to G^2

For scan-designed circuits (LSSD in particular), Goel has further projected the total test-application time in the following way.

For each test, the test-application time is proportional to the time to load/unload the scan register plus the actual time to drive and sense data on the driver/sensor pins. Generally speaking, the driver/sensor time is insignificant relative to the scan path load/unload time. Hence, test-application time is proportional to the load/unload time which, in turn, is proportional to the number of latches in the circuit. For large LSSD circuits, the total gate count G is usually not much more than the equivalent gate count of the latches. Hence we conclude, for each test, that the test-application time is itself proportional to G. Given that the total number of tests is also proportional to G, then the final result is as follows.

Result 5. The total test-application time for scan (LSSD) circuits is proportional to G^2

As we move forward to VLSI, defined here to mean devices with an equivalent gate count in excess of 100,000, then so the implications of Goel's results become disturbing. For example, if we compare a 1000–gate circuit with a 100,000–gate circuit, the numbers of tests will increase by a multiplicative factor of 100. However, test–generation time, deductive fault–simulation time and total test–application time will also each increase, this time by a multiplicative factor of 10,000 i.e. what was a test–generation time of 2 minutes, say, becomes a test–generation time of 333 hours i.e. 14 days.

In practice therefore, we should re–examine the statement that the test–generation problem is solved for combinational circuits. It is solved for circuits of 1000 or 10,000 gates but may not be for circuits of 100,000 gates or more. This means that we should not stop in our search for faster and more efficient test–generation techniques and, it is the author's opinion that techniques of the future will be more interactive than in the past and that they will be based on the concepts of expert systems employing knowledge items about the design and how it should be tested. An example of such a system – HITEST – has just been announced (references 6.6–6.8). Others are sure to follow.

As more intelligent test–generation systems, such as HITEST, are developed, then so testability analysis will take on a different aura. Currently, tools such as CAMELOT and SCOAP are based on a general methodology for generating tests. In reality, individual test–generators have their own personal characteristics and the next generation of testability–analysis tools will be based on a better understanding of the characteristics and limitations of the whole test environment i.e. not just the test–generator but also the fault simulator and the target ATE.

6.1.2 Test evaluation

The major method for evaluating the effectiveness of a set of stimuli patterns is to use a fault simulator, a modern example of which is the HILO simulator (reference 6.3). Fault simulation has limitations however: the cost of development is high; the problems of developing device models is rapidly becoming both time–consuming and complex; fault simulation time is unpredictable and often excessive for sequential designs and, even for combinational designs, is proportional to either G^2 or G^3 as discussed earlier.

The solutions will be twofold at least. One approach will be to continue to refine fault simulation techniques, such as the recently announced Parallel–Value–List (PVL) fault simulator (reference 6.5) used within the HITEST system. This simulator makes use of and extends the concepts of both parallel and concurrent (deductive) fault–simulation theory, and is a significant improvement in terms of implementation.

The alternative approach will be to explore other methods for qualifying the comprehensiveness of a set of tests without actually carrying out a full fault simulation – for example, by measuring the success of the test generator at applying standard tests to defined subcircuits within the circuit. An approach such as this would almost certainly be based on the use of knowledge items as described earlier in connection with future test-generation systems.

6.1.3 Test application

The development of automatic test equipment will continue with, probably, three major emphases: device testers, functional/system testers, individual designer's prototype testers. Device testers will need to keep apace with the requirements of high-speed VLSI devices, including the ability to apply routines to initiate and monitor self-test routines.

In the area of board testers, the increased use of leadless chip carriers and multi-layer boards will cause a swing away from in-circuit testing back to true edge-connector testing, simply because of the high density and limited access. This trend will reinforce the need for test-generation software that can handle circuits (device or board) at a functional level and with limited access.

The emergence of the engineering workstation will, eventually, create a need amongst designers for access to comprehensive test facilities either to assist in debugging design prototypes or to allow the development of prototype test programs. To be effective, these test facilities should be immediately available and they should become an integral part of the workstation. Such a tester will probably be a scaled-down version of the full production-line tester and will need to be accompanied by powerful and interactive test program preparation software.

6.2 CONCLUSIONS

The next decade will see some fundamental changes to the philosophy of testing and therefore to the concepts of testable designs. Perhaps the most exciting breakthrough will be in the area of test-generation software that, finally, accepts the need for intelligent human interaction and is designed to accomodate this need. The application of the concepts of expert systems will provide a rich source for innovation in this area, and will cause an outward shift in the boundaries of what constitutes a testable design. Testability is all about understanding the limits of the test environment. As this environment changes and expands, then so the ideas of testability will also change and expand. The challenge is to keep the two sets of ideas in line with each other: the keyword is 'Design-For-Test': the message is integration of design and test activities.

REFERENCES

6.1 Goel P., 1980, "Test generation costs analysis and projections", Proc. 17th IEEE Design Automatic Conf., pp 77–84.

6.2 Lineback J.R., 1981, "Self-testing processors cut costs", Electronics International, Vol.54, No 25, Dec.15, pp 110–112.

6.3 Rappaport A., 1983, "Digital-logic simulator software supports behavioural models", EDN, Vol.28, No. 10, May 12, pp 95–98.

6.4 Tulloss R.E., 1983, "Automated board testing: coping with complex circuits", IEEE Spectrum, Vol. 20, No. 7, July, pp 38–43.

6.5 Moorby P.M., 1983, "Fault simulation using parallel value lists", Proc. IEEE ICCAD Conf., Sept.

6.6 Wharton D.J., 1983, "The HITEST test generation system – overview", Proc. IEEE International Test Conf., Oct.

6.7 Robinson G.D., 1983, "HITEST – intelligent test generation", Proc. IEEE International Test Conf., Oct.

6.8 Maunder C.M., 1983, "HITEST test generation system – interfaces", Proc. IEEE International Test Conf., Oct.

6.9 Smith K., 1983, "Scan-path logic integrated on chip tests gate array", Electronics International, Vol. 56, No. 15, July 28, pp 85–86.

Appendix

AMD Am29818 Serial Shadow Register Device

As this book was going to press, Advanced Micro Devices announced the Am29818 Serial Shadow Register device. This device is the result of a three year joint development project between Advance Micro Devices and Monolithic Memories and both companies are joint holders of the patents. The Monolithic Memories device is the SN54/74S818 and, although internally different from the Am29818, both devices are interchangeable pin for pin.

The Am29818 device is the first off-the-shelf general purpose device with in-built scan-path facilities. To the author, it represents a significant step forward in terms of the ability of logic designers to produce scan-path designs. Also, AMD anticipate further devices, such as programmable logic arrays and gate arrays, that will utilise the serial shadow register concept.

Because of the importance of the device, the appendix reproduces parts of the Am29818 data sheet. This material is the copyright (1983) of Advanced Micro Devices Inc., and is reproduced here with the permission of the copyright owners, who also reserve all rights.

Further discussion on the Am29818 device can be found in:

Lee F., Coli V., and Miller W., 1983, "On-chip circuitry reveals system's logic states", Electronic Design, Vol. 31, No. 8, April 14, pp 199-124.

This issue of Electronic Design also contains editorial comment on the device (page 3).

154

AN INTRODUCTION TO SERIAL SHADOW REGISTER (SSR) DIAGNOSTICS

DIAGNOSTICS

A diagnostics capability provides the necessary functionality as well as a systematic method for detecting and pin-pointing hardware related failures in a system. This capability must be able to both *observe* intermediate test points and *control* intermediate signals — address, data, control and status — to exercise all portions of the system under test. These two capabilities, observability and controllability, provide the ability to establish a desired set of input conditions and state register values, sample the necessary outputs, and determine whether the system is functioning correctly.

TESTING COMBINATIONAL AND SEQUENTIAL NETWORKS

The problem of testing a combinational logic network is well understood (Figure 1). Sets of input signals (test vectors) are applied to the network and the network outputs are compared to the set of computed outputs (result vectors). In some cases sets of test vectors and result vectors can be generated in a computer-aided environment, minimizing engineering effort. Additionally, fault coverage analysis can be automated to provide a measure of how efficient a set of test vectors is at pin-pointing hardware failures. For example, a popular measure of fault coverage computes the percentage of stuck-at-ones (nodes with outputs always HIGH) and stuck-at-zeros (nodes with outputs always LOW) a given set of test vectors will discover.

A sequential network (Figure 2) is much more difficult to test systematically. The outputs of a sequential network depend not only on the present inputs but also on the internal state of the network. Initializing the internal state register to the value necessary to test a given set of inputs is difficult at best, and not easily automated. Additionally, observing the internal state of a sequential network can be very difficult and time consuming if the state information is not directly available. For example, consider the problem of determining the value of an internal 16-bit counter if only a carry-out signal is available. The counter must be clocked until it reaches the carry-out state and the starting value computed. Up to 65,535 clock cycles may be necessary! An easier method must exist. Serial Shadow Register diagnostics provides this method.

SERIAL SHADOW REGISTER DIAGNOSTICS

Serial Shadow Register diagnostics provides sufficient observability and controllability to turn any sequential network into a combinational network. This is accomplished by providing the means to both initialize (control) and sample (observe) the state elements of a sequential network. Figure 3 shows the method by which serial shadow register diagnostics accomplishes these two functions.

Serial Shadow Register diagnostics utilizes an extra multiplexer on the input of each state register and a duplicate or shadow of each state flip/flop in an additional register. The shadow register can be loaded serially via the serial data input (thus the name

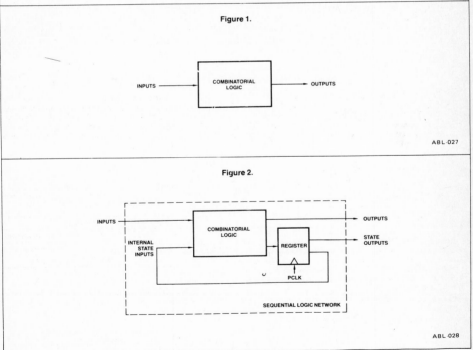

Figure 1.

ABL-027

Figure 2.

SEQUENTIAL LOGIC NETWORK

ABL-028

8

Figure 3.

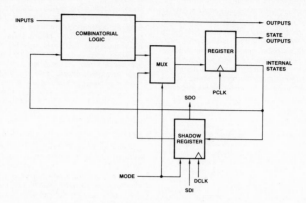

ABL-029

Figure 4.

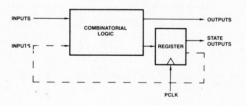

ABL-030

Serial Shadow Register diagnostics) for controllability. Once the desired state information is loaded into the serial register it can be transferred into the internal state register by selecting the multiplexer and clocking the state register with PLCK. This allows any internal state to be set to a desired state in a simple, quick, and systematic manner.

Internal state information can be sampled by loading the serial register from the state register outputs. This state information can then be shifted out via the serial data output to provide observability. Notice that the serial data inputs and outputs can

be cascaded to make long chains of state information available on a minimum number of connections.

In effect, Serial Shadow Register diagnostics breaks the normal feedback path of the sequential network and establishes a logical path with which inputs can be defined and outputs sampled (Figure 4). This means that those techniques which have been developed to test combinational networks can be applied to any sequential network in which Serial Shadow Register diagnostics is utilized.

A TYPICAL COMPUTER ARCHITECTURE WITH SSR DIAGNOSTICS

When normal pipeline registers are replaced by SSR diagnostics pipeline registers system debug and diagnostics are easily implemented. State information which was inaccessible is now both observable and controllable. Figure 5 shows a typical computer system using the Am29818.

Serial paths have been added to all the important state registers (macro instruction, data, status, address, and micro instruction registers). This extra path will make it easier to diagnose system failures by breaking the feed-back paths and turning sequential state machines into combinatorial logic blocks. For example, the status outputs of the ALU may be checked by loading the micro instruction register with the necessary micro instruction. The desired ALU function is then executed and the status outputs captured in the status register. The status bits can then be serially shifted out and checked for validity.

A single diagnostic loop was shown in Figure 5 for simplicity, but several loops can be employed in more complicated systems to reduce scan time. Additionally, the Am29818's can be used to sample intermediate test points not associated with normal state information. These additional test points can further ease diagnostics, testability and debug.

Figure 5. Typical System Configuration

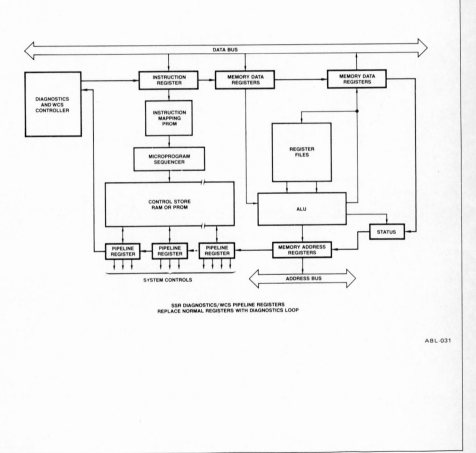

SSR DIAGNOSTICS/WCS PIPELINE REGISTERS
REPLACE NORMAL REGISTERS WITH DIAGNOSTICS LOOP

ABL-031

10

USE OF THE Am29818 PIPELINE REGISTER IN WRITABLE CONTROL STORE (WCS) DESIGNS

The Am29818 SSR diagnostics/WCS Pipeline Register was designed specifically to support writable control store designs. In the past, designers of WCS based systems needed to use an excessive amount of support circuitry to implement a WCS. As shown in Figure 7 additional input and output buffers are necessary to provide paths from the parallel input data bus to the memory, and from the instruction register to the output data bus. The input port is necessary to write data to the control store, initializing the micromemory. The output port provides the access to the instruction register, indirectly allowing the RAM to be read. Additionally, access to the instruction register is useful during system debugging and system diagnostics.

The Am29818 supports all of the above operations (and more) without any support circuitry. Figure 6 shows a typical WCS design with the Am29818. Access to memory is now possible over the serial diagnostics port. The instruction register contents may be read by serially shifting the information out on the diagnostics port. Additionally, the instruction register may be written from the serial port via the shadow register. This simplifies system debug and diagnostics operations considerably.

CONCLUSION

Serial Shadow Register diagnostics provides the observability and controllability necessary to take any sequential network and turn it into a combinational network. This provides a method for pin-pointing digital system hardware failures in a systematic and well understood fashion.

Figure 6. Am29818 Based WCS Application.

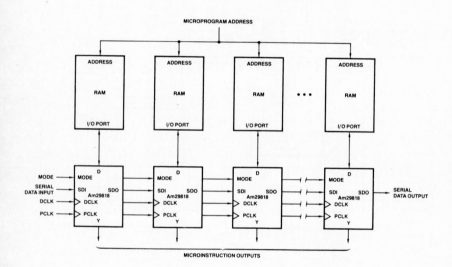

ABL-032

11

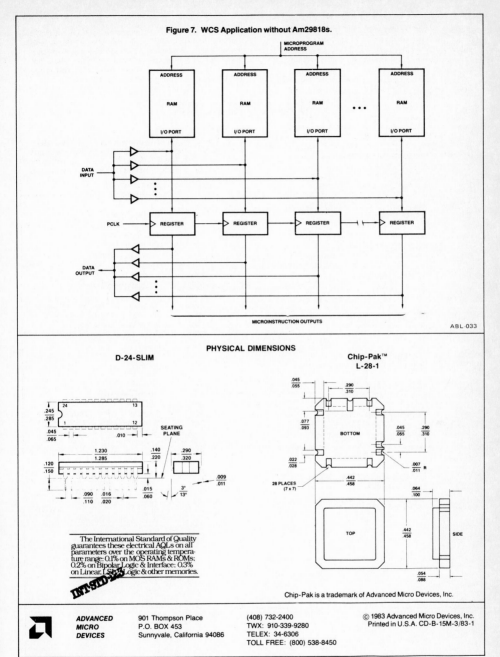

Figure 7. WCS Application without Am29818s.

ABL-033

PHYSICAL DIMENSIONS

D-24-SLIM

Chip-Pak™ L-28-1

The International Standard of Quality guarantees these electrical AQLs on all parameters over the operating temperature range: 0.1% on MOS RAMs & ROMs; 0.2% on Bipolar Logic & Interface; 0.3% on Linear, LSI Logic & other memories.

Chip-Pak is a trademark of Advanced Micro Devices, Inc.

ADVANCED MICRO DEVICES
901 Thompson Place
P.O. BOX 453
Sunnyvale, California 94086

(408) 732-2400
TWX: 910-339-9280
TELEX: 34-6306
TOLL FREE: (800) 538-8450

© 1983 Advanced Micro Devices, Inc.
Printed in U.S.A. CD-B-15M-3/83-1

Abbreviations

ATE	Automatic Test Equipment
BILBO	Built-In Logic Block Observer
BUT	Board-Under-Test
CAMELOT	Computer-Aided MEasure for LOgic Testability
cpu	central processing unit
CRC	Cyclic-Redundancy Check
CTF	Controllability Transfer Factor
CY	Controllability
DEMUX	Demultiplexer
DFL	Detected Fault List
DFT	Design-For-Testability
DIP	Dual Inline Package
DUT	Device-Under-Test
ECL	Emitter Coupled Logic
ECO	Engineering Change Order
KGB	Known Good Board
KGD	Known Good Device
LED	Light-Emitting Diode
LFSR	Linear Feedback Shift Register
LSI	Large Scale Integration
LSSD	Level-Sensitive Scan Design
MOS	Metal-Oxide Silicon
MSI	Medium Scale Integration
MUX	Multiplexer
NPDC	Non-Propagating D-Cube
OB	Objective

OTF	Observability Transfer Factor
OY	Observability
PAL	Programmable Array Logic
PCB	Printed-Circuit Board
PDC	Propagating D-Cube
PI	Primary Input
PO	Primary Output
PODEM	Path Oriented DEcision Making
p-r	pseudo-random
PVL	Parallel-Value-List
RAM	Random Access Memory
RAPS	RAndom Path Sensitising
RAS	Random Access Scan
ROM	Read Only Memory
s-a-1	Stuck-at-1
s-a-0	Stuck-at-0
SA	Signature Analysis
SDI	Scan Data In
SDO	Scan Data Out
SFP	Single Fault Propagation
SOT	Set Of Tests
S-REG	Shift Register
SRL	Shift Register Latch
SSI	Small Scale Integration
TFL	Target Fault List
TNC	Terminal Numbering Convention
TTL	Transistor Transistor Logic
TY	Testability
VLSI	Very Large Scale Integration

Index

monostable, 130
multi-path D-drive, 96

node, 84
node excitation, 69
non-propagating D-cube, 17

objective, 86
observability, 16, 115
Observability Transfer
  Factor, 17
open-collector outputs, 133
oscillator, 135
overdrive, 116

p-r generator, 72
PAL devices, 132, 154
parallel simulation, 111
Parallel-Value-List, 151
partition, 123
PI-remake, 96
PODEM, 83, 89
predictability, 40
primary input, 2
primary output, 2
propagating D-cube, 17
pseudo-random, 69
pulse catching, 142

race, 50, 127
Random Access Scan, 50
RAPS, 83, 98
reconvergence, 23, 96
redundancy, 41, 121
ROM devices, 132

Scan advantages, 60
Scan design, 46, 59
Scan path, 47
Scan penalties, 60
Scan Set, 46, 120
Scan test sequence, 75
Schneider Counterexample,
  98

SCOAP, 40
self-test, 70
serial simulation, 111
shadow register, 121, 154
Shift Register Latch, 53
shift test, 49
signature, 63
Signature Analysis, 67
Single Fault Propagation,
  111
single-latch, 55
singular component, 132
SN54/74S818, 154
static test compaction, 106
stay-alive circuit, 142
stuck-at-0, 2
stuck-at-1, 2

Terminal Numbering
  Convention, 89, 113
test, 2
test application, 4, 62
test costs, 6, 150
test evaluation, 3, 61
test generation, 3, 61,
  115
test pattern, 2
test point, 31, 116
test-generation strategy,
  35
test-state register, 120
TEST/80, 39
testability, 24
testable (definition), 6
TESTSCREEN, 40
timing analysis, 61
timing faults, 60
TMEAS, 38
tristate outputs, 133

VICTOR, 41

wired-AND, 136
wired-OR, 136